Praise for
Runaway Climate

Rapid, catastrophic climate change may have little precedent in human history, but the rocks tell us it has happened before. This riveting book clearly outlines the potential scope of the crisis that we are unleashing through our continued burning of fossil fuels. If you care at all about our future, you must read *Runaway Climate*.

—Richard Heinberg, Senior Fellow, Post Carbon Institute
and author, *Power: Limits and Prospects for Human Survival*

Climatologists provided us with early warning of the climate crisis, and now—as this fascinating account makes clear—geologists are making clear that the past both confirms those warnings and intensifies them. Reading this will, I hope, be a prelude to activism that matters.

—Bill McKibben, author, *The End of Nature*

Earle's new book is a compelling call to climate action that is uniquely engaging and disturbing in equal measure. By setting today's climate crisis within the long story of our planet, he invites us all to acknowledge realities of these times and to find inspiration to act in the climate solutions stories that he shares.

–Laura Lengnick, author, *Resilient Agriculture:*
Cultivating Food Systems for a Changing Climate

I love it. Earle understands the big climate picture and paints it with exceptional clarity.

—James Hansen, director, Climate Science, Awareness and Solutions,
Columbia University Earth Institute

An Informative, succinct, and fascinating read—Steven Earle offers a unique and detailed account of Earth's climate history. His innate story-telling ability, coupled with his remarkable talent for making complex scientific information accessible, makes this page-turner a must read for anyone seeking to understand the Earth's climate system.

—Andrew Weaver, Professor, University of Victoria, former Lead Author, Intergovernmental Panel on Climate Change 2nd, 3rd, 4th and 5th Scientific Assessments, former chief editor, the *Journal of Climate*

An engaging tour through the complex natural processes at play in writing the Earth's long history of natural climate change to our present climate emergency. This primer will give campaigners, policymakers, and concerned citizens a more thorough understanding of climate science and renewed conviction to go all in on applying the brakes, leaving fossil fuels behind, and embracing a cleaner, healthier, and more equitable future.

—Tom Green, Senior Climate Policy Advisor, David Suzuki Foundation

RUNAWAY CLIMATE

RUNAWAY
CLIMATE

WHAT THE GEOLOGICAL PAST CAN TELL US ABOUT THE COMING CLIMATE CHANGE CATASTROPHE

Steven Earle, PhD

new society
PUBLISHERS

Cover design by Diane McIntosh.
Cover image: Steven Earle, PETM-aged sedimentary rocks at Polecat
Bench, Bighorn Basin, Wyoming. Box background: Adobestock_86345603.

Printed in Canada. First printing April 2024.

Inquiries regarding requests to reprint all or part of *Runaway Climate* should be addressed to New Society Publishers at the address below. To order directly from the publishers, please call 250-247-9737 or order online at www.newsociety.com. Any other inquiries can be directed by mail to:

New Society Publishers
P.O. Box 189, Gabriola Island, BC V0R 1X0, Canada
(250) 247-9737

LIBRARY AND ARCHIVES CANADA CATALOGUING IN PUBLICATION
Title: Runaway climate : what the geological past can tell us about the
    coming climate change catastrophe / Steven Earle, PhD.
Names: Earle, Steven, author.
Description: Includes bibliographical references and index.
Identifiers: Canadiana (print) 20230585787 | Canadiana (ebook)
    20230585833 | ISBN 9780865719897 (softcover) |
    ISBN 9781771423786 (EPUB) | ISBN 9781550927825 (PDF)
Subjects: LCSH: Climatic changes—History. | LCSH: Paleoclimatology.
    | LCSH: Paleontology—Paleocene. | LCSH: Paleontology—Eocene. |
    LCSH: Climatic changes—Forecasting.
Classification: LCC QC903 .E276 2024 | DDC 551.609—dc23

New Society Publishers' mission is to publish books that contribute in fundamental ways to building an ecologically sustainable and just society, and to do so with the least possible impact on the environment, in a manner that models this vision.

Synopsis

T HE PALEOCENE-EOCENE THERMAL MAXIMUM (PETM) started 56 million years ago and lasted for about 180 thousand years. The Earth's air and water temperatures rose by 5° to 8°C, and the climate became stormier but drier overall. The oceans were acidified and depleted in oxygen. About half of deep-ocean foraminifera went extinct, and coral reefs were decimated. Land vegetation changed dramatically as climate zones shifted towards the poles. Most land animals were forced to migrate to new locations. A few became extinct, and some got a lot smaller. The early part of the PETM was associated with the first appearance of artiodactyls, perissodactyls (even- and odd-toed hoofed animals respectively), and primates.

PETM may have been triggered by volcanism in the northern Atlantic or by Milankovitch cycles, or both, but it is likely that other strong climate feedbacks, such as release of methane through breakdown of permafrost and organic matter, and eventually the destabilization of deep-ocean methane hydrates, were the real drivers of change. The Earth's temperature started to gradually decrease back to "normal" after about 100,000 years, likely because of negative feedbacks such as enhanced weathering of rocks and enhanced ocean productivity.

We know that our civilization will be seriously challenged by the climate change we can already see happening, and that this will make vast areas of our planet uninhabitable and create an unimaginable refugee situation. What the PETM teaches us is that there is a real possibility that anthropogenic climate change could push us into runaway climate change that would be an order of magnitude worse still. We are emitting as much carbon as was emitted back then, but we're doing it far faster (which gives it more impact), and the strong feedbacks associated with rapidly melting sea ice and glacial ice (which did not exist then) are accelerating the warming. The implications for our civilization are beyond disastrous. The combination of sea-level rise, supercharged storms, excessive heat, and intense aridity would make all of us refugees, most of us hungry and thirsty, and very many of us dead.

Although nobody knows whether we are on track for a PETM-like future, we do know that we need to do something about regular climate change, and recognizing the credible risk of a crisis many times worse, it's obvious that we must act decisively and quickly. We need to dramatically reduce our use of fossil fuels now and end it altogether within two decades. That will take significant individual changes and very strong government policies.

We got ourselves into this mess, and we can get ourselves out! The consequences of not doing so are frightening. There isn't much time.

Contents

To my grandchildren, Elliot, Bennie, and Jack,
with love and with apologies for what we have done.

Acknowledgments

I SINCERELY THANK Rob West of New Society Publishers for trusting me to tell this important story and for his informed and solid advice throughout the process. It has made all the difference. Thanks to everyone else at NSP for your enthusiasm and help.

I am especially indebted to Isaac Earle and Kathleen Maser for reading every word and providing serious and constructive criticism and encouragement. Finally, I could not have done this without support from my family, especially Justine, Kate, Rosie, Heather, and Tim.

I am grateful to be able to live on the unceded territory of the Snuneymuxw First Nation, who have taken good care of this land and the surrounding waters since time immemorial.

Preface

If you don't change direction, you may end up where you are heading.

—Lao Tzu

WE ARE AT A TURNING POINT. Yes, of course we are always at turning points, both figuratively and literally, but this is an existential one, and we have to decide whether or not to change direction. Do we need to make a life-changing sharp turn towards reducing our impact on the climate system? Can we afford a much slower and less disruptive turn? Or should we not turn at all, and just continue blithely in our present direction?

Climate science insists that we turn. Now! It tells us clearly and unequivocally that there is something very dangerous up ahead that needs our focussed attention. If we turn sharply, we have a chance of staying out of trouble. If we don't, we are likely to find that the road ahead is so potholed and muddy that we are brought to a complete halt. We will all have to get out and push, but the mud may be too deep. We'll be walking the rest of the way; I hope you have practical shoes.

It's also possible that this road takes us into unknown territory, to a place where it becomes unclear if there is still a viable way ahead, where the visibility is poor and getting worse, where the grade steepens, and the surface becomes slipperier. No this isn't a road any more, not even a track! Pumping the brakes makes no difference at all. Skidding, swerving, and screaming, there is nothing we can do to stop our slow, sickening slide over the edge. That brings this particular journey to an end, even if we don't all perish.

Climate science may not know which of these scenarios is in store for us, but it does know that our future road is unlike anything we humans have ever been on. *Runaway Climate* is about a well-documented example of the scenario where we all scream as we go over a cliff. At the end of the Paleocene, 56 million years ago, life on Earth was gobsmacked by runaway climate change, an event known as the Paleocene Eocene Thermal Maximum or PETM. The organisms that we now share the planet with, including our own distant ancestors, obviously did survive the PETM, but the world changed so much that the type of civilization that we have today could not have survived, and likely could not have recovered for at least 100,000 years. The premise here is that devastating climate change, like that which took place 56 million years ago, could happen again because of how we are changing the climate, and there are good reasons to believe that the changes that could be in store for us would be much faster, and therefore significantly more devastating, than what happened back then.

Anyone who is not asleep at the wheel can see that the road we're on is already problematic. It is bumpy, unpredictable, and difficult to negotiate. It is getting progressively worse, even if it isn't yet obvious that we're about to become mired in mud or slide over a cliff. If we are fully confident that the way ahead is going to be smooth and straight once we're through this rough patch, then it is understandable that we might decide to just carry on and hope for the best. But if we just suspect that there could be serious climate danger ahead—and that it could be far worse than the climate change that we already know is coming—we need to slow right down, pull over to the side, and stop. We then need to think carefully about our options, and find a new course. Most of the changes we need to make will have multiple positive side benefits.

—Steven Earle, August 2023

Part I

What Happened in the Past?

This book is about a geological event that took place 56 million years ago, when the Earth's climate warmed dramatically over as little as a thousand years, stayed dangerously warm for about 180,000 years, and then cooled again. It's known as the Paleocene Eocene Thermal Maximum (PETM) because it happened at the boundary between the Paleocene and the Eocene Epochs. In the first five chapters, we will examine the geological evidence for the PETM and discuss how the oceans and the land were affected. In chapters 6 and 7, we will delve into the likely mechanisms for the dramatic climate change of the PETM and then assess whether or not the Earth's current geological and biological state, and the exceptionally fast climate change that we are causing, could lead us into a similar crisis.

Part II includes a description of how extraordinarily difficult a future PETM-like climate would be for human civilization and for most of the other inhabitants of the Earth, and a discussion of what immediate action governments, corporations, and individuals need to take to avoid that fate.

Unless otherwise noted, all drawings and photos are by the author.

Chapter 1

The Bighorn Basin

It is certainly within the domain of science to determine when the earth was first fitted to receive life, and in what form the earliest life began. To trace that life in its manifold changes through past ages to the present is a more difficult task, but one from which modern science does not shrink.

—Othniel Marsh, 1877[1]

THE BIGHORN BASIN is a thumbprint-shaped pocket of northwestern Wyoming about the size of Lake Ontario or Connecticut or Northern Ireland (figure 1.1). It consists of dry brown rolling hills and green flat river valleys and is surrounded by mountain ranges: the Bighorns to the east, the Absarokas (and Yellowstone National Park) to the west, and the Owl

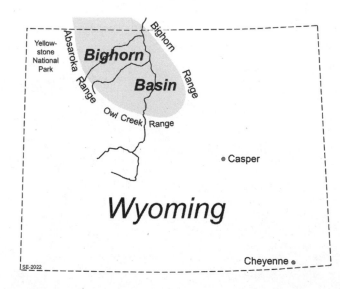

Figure 1.1:
The location of the Bighorn Basin in northwestern Wyoming.

3

Creek Range to the south. It can be uncomfortably hot in the summer, and cold and snowy in the winter. Although it is a semidesert and dominated by sagebrush and grasses, it is well watered by streams. The main one is the Bighorn River, which flows in from the south through a gap in the Owl Creek Range, and out to the north through a gap in the Bighorn Range and then on to join the Yellowstone River.

The basin was originally home to the Eastern Shoshone People. Chief Washakie ("Shoots the Buffalo Running"), a warrior and diplomat, was prominent among the Eastern Shoshone in the nineteenth century. Shoshone territory was colonized by ranchers and was made famous by colourful characters like Buffalo Bill Cody and Butch Cassidy, and by ambitious fossil hunters such as Edward Cope and Othniel Marsh. Today the basin is an important farming region—where irrigation water is available—and it is dotted with oil and gas wells. But it is still only sparsely populated, with no town exceeding 10,000.

So, what are we doing here in the Bighorn? As you might have guessed, it's not the sagebrush, the pronghorn antelope, or the wall-to-wall Republicans that we're here to observe, but what's underneath, and there is a fascinating geological history that extends over 500 million years. This history is written in the rocks shown in cross-section in figure 1.2. The oldest rocks of Wyoming are Precambrian in age (older than 539 million years), and they are a southern extension of the metamorphic and igneous rocks of the Canadian Shield. These are overlain by a series of sedimentary layers ranging in age from about 500 million (Cambrian) to around 50 million years (Eocene). Unlike the metamorphosed Precambrian rocks underneath, these ones have lots of great

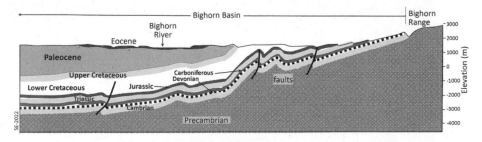

Figure 1.2: *Geological cross-section of the eastern part of the Bighorn Basin. (Based on Wyoming State Geological Survey, Geological Cross Sections, Bighorn Basin.)*

fossils, and they tell a wonderful story about the geological history of North America.

That's interesting, but in fact we're really here because this small part of North America has arguably the best rocks in the world for studying the short interval at the start of Eocene (56 million years ago) known as the Paleocene-Eocene Thermal Maximum (PETM), a short period of runaway climate change that we all might benefit from knowing more about.

Before we go back in time, it would be useful to quickly review the geological time scale. The last 600 million years are represented in figure 1.3 (although this is only about one-eighth of all of the Earth's history). The diagram is labelled with the approximate times of some of the important events in the history of life on Earth.

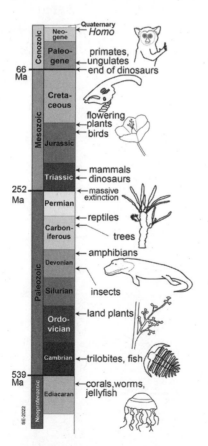

Figure 1.3: *The geological time scale for the past 600 million years. (Based on the International Commission on Stratigraphy, stratigraphy.org.)*

The oldest sedimentary rocks in the Bighorn region[2] are Cambrian in age—about 500 Ma (Ma = mega-annum, so that's 500 million years ago)—and they are dominated by limestone, formed in warm shallow water off the coast of the continent of Laurentia, which straddled the equator at that time (figure 1.4). Laurentia forms the core of what is now North America. There was no actual Bighorn Basin then. The Maurice Limestone, which has fossils of trilobites and brachiopods, is contemporaneous with and formed in a similar continent-margin setting to British Columbia's Burgess Shale, which has abundant evidence of what we call the Cambrian Explosion: the rapid evolution of the ancestors of many of Earth's modern life-forms.

About 100 million years later, in the early Devonian, Laurentia had moved farther south and the area that is now the Bighorn was right on the coast (figure 1.4). There were small primitive plants on land at this time (e.g., *Psilophyton*,[3] a plant without leaves or roots) and arthropods (e.g., scorpion-like creatures). A two-metre-long aquatic arthropod

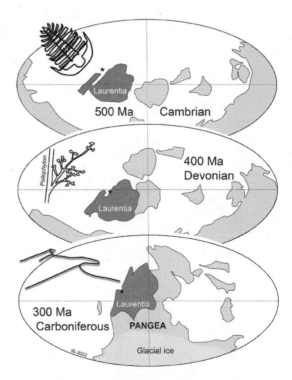

Figure 1.4: *Distribution of the continents in the Cambrian, Devonian, and Carboniferous. The star indicates the approximate location of the Bighorn Basin. (Based on maps by Christopher Scotese.[4])*

known as eurypterid patrolled the shallow water hunting for fish and other creatures.

Another 100 million years later, almost all the continental areas have been pushed together by plate tectonics, creating one super-continent known as Pangea. Land plants had become large and abundant by the early Carboniferous, and their vigorous growth consumed enough of the atmosphere's carbon dioxide to cool the climate. That allowed for accumulation of snow in the southern polar regions and then to the formation of glaciers, which eventually covered the whole of southern Pangea. There was no glacial ice in Laurentia, but the Bighorn area was a cold and windy desert with massive sand dunes. Bighorn Basin rocks from this time have few fossils, but plants were still abundant in other parts of the world and vertebrates (amphibians and reptiles) had started to colonize the land.

Pangea was still mostly in one piece at around 220 Ma, during the Triassic (figure 1.5), and was slowly moving north. Small continents

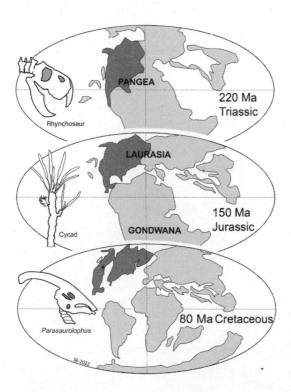

Figure 1.5: *Distribution of the continents during the Triassic, Jurassic, and Cretaceous.*

collided with and got stuck onto its western edge so that the Bighorn region was now much further inland. Sedimentary rocks from this time have a distinctive red colour because much of the area was a hot desert. The Bighorn Basin's Triassic Chugwater Formation has fossils of rhynchosaurs, an herbivorous reptile with a parrot-like "beak" and lots of teeth for grinding up vegetation. Dinosaurs did exist by this time (in what are now South America and Africa), but they are not known in the Triassic rocks of the Bighorn.

By the late Jurassic (150 Ma), Pangea had started to split apart along a line between Laurentia and the "Africa" part of Gondwana, forming the proto-Atlantic Ocean. The Jurassic rocks of the Bighorn Basin include the Morrison Formation, arguably the most prolific sequence of rocks for dinosaur fossils in the world, with skeletons of giant long-necked sauropods, bizarre ornithopods, spiky ankylosaurs, and fierce theropods both large and small, but no birds yet (at least not here). There are also fossils of various small mammals and reptiles. The region was richly forested with conifers, cycads, and ginkgoes, and with abundant ferns and horsetails in the understory, but no flowering plants yet.

The Cretaceous world of 80 Ma is starting to look more like the modern world. Pangea has split apart, and North and South America, Africa, and Eurasia have become separate continents. India is steaming north towards Asia, but Australia is still attached to Antarctica. It was warm almost everywhere, so sea level was high because there were no glaciers. Part of the interior of North America had been pulled down-ward by an underlying subducting oceanic plate, creating a wide inland sea. Giant ammonites lurked in the deeper water, along with large marine reptiles. Dinosaurs (such as *Parasaurolophus*) were still the dominant terrestrial animals. Pterosaurs—which are not dinosaurs—patrolled the skies. The forests were not very different from those of the Jurassic, except that angiosperms (flowering plants) were present, both as trees and shrubs, and there were birds.

Kapow!

Almost everything changed in an instant at 66 Ma, when a 12-kilo-metere diameter meteoroid slammed into Yucatan, Mexico, marking the end of the Cretaceous and the Mesozoic and the beginning of the

Paleogene and the Cenozoic. A massive volume of debris was blasted out of the crust, and as the fragments glowed white hot on re-entry, they generated flesh-burning heat over much of the Earth and sparked continent-scale wildfires. That was followed by months of near-complete darkness and bitter cold, and then by strong warming. Gone forever were terrestrial dinosaurs, pterosaurs, ammonites, and giant marine reptiles. Most bird and mammal groups survived, and once things settled down, most of the plants that had thrived in the late Cretaceous regenerated from seeds and roots.

Early in the Paleogene, the continents were close to their current locations (figure 1.6), but the northern part of the Atlantic may still have been closed between Greenland and Europe. India had not quite converged with Asia. It was very warm almost everywhere. The Fort Union Formation shale of the Bighorn Basin provides evidence that forested swampy conditions were common, and those swamps were home to small mammals, living in the trees (like *Plesiadapsis*), and to crocodiles and turtles.

The part of the geological time scale that is most relevant to this book is shown on figure 1.7. The Paleocene and Eocene are epochs within the Paleogene Period.[5] By the late Paleocene, the Earth's average temperature was a few degrees warmer than it was during the Cretaceous and about 10°C warmer than it is today. There were no glaciers anywhere,

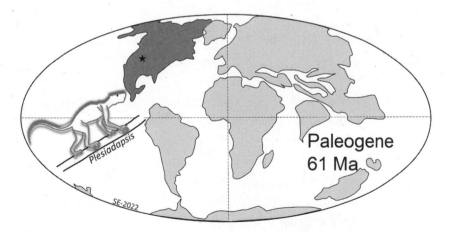

Figure 1.6: *Distribution of the continents during the Paleogene.*

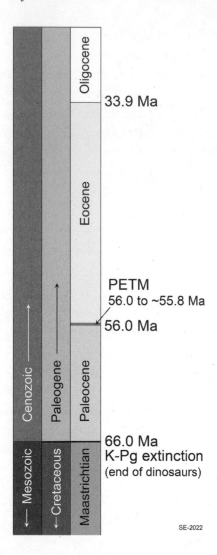

Figure 1.7: *The geological time scale for the early part of the Cenozoic. (Based on the International Commission on Stratigraphy, stratigraphy.org)*

although there was likely extensive permafrost and snow in Antarctica and some mountainous regions (but not the Himalayas, as they didn't exist yet). Because no water was tied up in glacial ice, sea level was at least 70 meters higher than it is today. Plants and animals of the time were able to cope with high temperatures because they had evolved in hot climates, and because natural climate change is typically very slow. Over the 10 million years of the Paleocene, the climate warmed by a total of about 3°C.[6] Late Paleocene plant communities were not very different from those of the Cretaceous, and also not that different from modern-day communities.

Late Paleocene in the Bighorn Basin

Let's pretend that we can time-travel to a Bighorn Basin forest at the very end of the Paleocene. Dial up "56.0 Ma" on your time machine, and you might find yourself in the welcome shade of some cypress conifers, with a few metasequoias in the mix. In wetter parts of the region, you will be able to wander through birch forests where you should also see walnut, laurel, magnolia, katsura, and dogwood. The understory will be populated with ferns and horsetails, along with ancestors of modern grapes and maybe even some grasses. In other words, many of the late Paleocene plants will be recognizable to most people. We all might feel at home in such a forest, if not for the stifling heat.

Be still now. Listen to the wind in the trees and the sounds of strange creatures. If you wait long enough, you might get lucky and see some bizarre mammals with unpronounceable names. The most common are the Eulipotyphla (means "truly fat and blind") insectivores, which are the ancestors of modern-day shrews, moles, and hedgehogs, but you'll have to look carefully, as these weigh in at just a few grams and they are good at hiding. Next most common are the Multituberculata, which are a little bigger, some up to nearly a kilogram. They are rodent-like (but not actual rodents) and have no living descendants, so will appear quite strange. Up in the trees, you might catch a glimpse of a Primatomorpha. They are mostly small, but some could be up to a few kilograms. The most common in this forest is *Plesiadapis,* which is about the size of a weasel. True primates were not present during the Paleocene. This one is *not* your distant ancestor, but it may be within a sister clade to the ancestors of the real primates that evolved early in the Eocene. If you hear something larger cracking a dry twig, it could be one of the condylarths. Again, there is a lack of consensus about this group, but they are considered to be the ancestors of hoofed animals (ungulates). The biggest ones around here are *Phenacodus*, at around 50 kilograms, and *Ectocion* at around 10 kilograms. Both are thought, by some, to be ancestors to the perissodactyls, ungulates with odd numbers of weight-bearing toes (1 or 3), like horses. If you hear something bigger still, off in the distance, that could be a Dinocerata. This is another hoofed mammal, not a dinosaur. Most were the size of pigs (around 50 kilograms), but they could reach rhino-size, so you might want to look for a tree to hide behind, or climb! If one of them comes into view, it will be like nothing you've ever seen (or hope to see!), with an array of tusk-like nobs protruding upward and outward from its skull and its snout and two sabre-like teeth extending down from the front of the upper jaw. Don't panic, it is an herbivore, and does not want to eat you! Dinoceratas disappeared in the Oligocene. And finally, while you're not likely to get this lucky, there should be some representatives of the order Carnivora in this forest, including *Didymictus*, which weighs in at around 5 kilograms (about the size of a small fox). *Didymictus* and its relations are more accurately carnivoramorphs, because while they have carnivore-like teeth, they lack other important carnivore features. They are not the ancestors of your cat or your dog.[7]

Most of the modern orders of birds (excluding the Passeriformes, or perching birds) had evolved by this time,[8] but bird fossils are rare in the Paleocene rocks of Wyoming, probably because the delicate bones of birds don't fossilize well. Nevertheless, it is likely that some of the calls that you hear in this forest are from birds, and those would probably be ancestors of the raptors, woodpeckers, or maybe even ducks or geese. You won't hear any bird calls that are "musical," at least not to your ears.

Earliest Eocene

Now, let's tick forward in time by just 10,000 years, into the early Eocene. Turn the dial to 55.99 Ma. The floodplain shale beds of the Eocene Willwood Formation are of primary importance to us here. They are mostly drab grays and browns, but the lowest 60 meters of the formation (representing the earliest Eocene) has distinctive fiery red and dull purple layers. These are from a critical interval in Earth history known as the Paleocene-Eocene Thermal Maximum (PETM), and our goal (mine, and yours I hope) in this book is to try to understand what happened, how and why it happened, whether it could happen again, and finally, what we can do to prevent that.

Evidence from many locations around the world (presented in later chapters) shows that the Earth's average temperature—which, as you know, was already hot by today's standards—soared by another 6° to 8°C over a period of a few thousand years (or less) at the start of the Eocene, and that there were other dramatic changes, both on land and in the ocean. While that rate of change was still much slower (about 10 times slower) than what humans have caused over the past century, it was fast enough to result in significant changes to the plant and animal communities of the Bighorn Basin. Some of those changes had implications for us humans and for the animal communities that we coexist with now.

At 55.99 Ma in the PETM, the early Eocene forest is completely different from the one you wandered through at the end of the Paleocene. The conifers are gone, as are most of the other plants that might have looked familiar earlier; in fact, over 90% of the plant species that were here just 10,000 years ago are gone. The predominant tree species now is a legume that is related to the modern-day mimosa (sorry, the tree, not the drink, aka, silk tree or *Albizia*). These types of plants were not present in this region prior to the PETM but had been growing over 1,000 kilometers further south. This forest is also much brighter than the one that was here before, and where there were swamps and rivers before, now there are just ephemeral streams.

Let's be quiet now and see what shows up. There's a chance that we'll get a fleeting view of a tiny *Diacodexis*, which could be the very earliest of the artiodactyls (hoofed animals with an even number of weight-bearing toes). It's under 3 kilograms and only about 50 centimeters long, with a tail almost as long. *Diacodexis's* relations are the ancestors of deer, antelope, goats, sheep, moose, giraffes, and whales. Yes, whales! This *Diacodexis* may have been in a hurry to avoid meeting up with something known as *Arfia*, which is a type of carnivore, but not like any modern ones. *Arfia* is in the Hyeanadontidae family. That just means that they have "hyena-like teeth"; they are not related to the hyenas that you're thinking of. Listen, something small is moving around in the tree above you. If you're lucky, you might catch a glimpse of a *Teilhardina*. It will look like a marmoset or a tarsier but is no bigger than a mouse

Figure 1.8: Sifrhippus, *the earliest known horse, with the lower leg of a modern-day horse for scale. (Based on an Eduard Solà photograph of a specimen in the Swedish Museum of Natural History.[9])*

and is indeed the earliest known primate. Yes, this one actually could be your distant ancestor. They might have originated in southern Asia, and if so, it likely took many hundreds of generations for them to reach the Bighorn Basin. *Teilhardina* leaps to another branch, and another, and then it disappears behind the leaves.

All is quiet for a time, but then something else catches your eye within a dry stream channel. No? Nothing there? Wait, yes, there is something there, but it must be small. Oh, it's our lucky day! That tiny thing is a *Sifrhippus*, which means "Zero horse," because, as far as we know, this is the first ever horse (figure 1.8). It's no bigger than a fox! It nibbles on some leaves and some blades of grass. It flicks its tail to ward off a fly, and then twitches one of its ears. It's looking nervous, which isn't surprising for a horse, especially one this size. It's probably not alone, but its companions are not in sight.

It's very likely that *Sifrhippus* evolved here in North America,[10] in the earliest Eocene. There are early Eocene equids in western Europe, but they are in rocks that are just a little bit younger than those of the

Bighorn. Note that it has three toes, making it a perissodactyl (odd number of weight-bearing toes). Over the next several million years, two of those toes will get shorter as horses evolve to have only one weight-bearing toe, like the modern horse foot beside it.

Where did it go? It must have heard, or smelt, or felt something that we didn't. Oh shit! What was that awful shriek? There's only one thing in this forest that could make a noise like that, *Gastornis*! It's a giant flightless bird, about two meters tall, with an oversized head and massive beak. Although there is some debate about what *Gastornis* ate, the recent evidence points to a vegetarian diet, so we and *Sifrhippus* can relax a little. What a day! A ridiculously small horse and a ridiculously large bird, all in a matter of minutes. But wait, if *Gastornis* isn't a predator, it must be something's prey, and so why did it make such a fearful noise?

Yikes! Enough time travel for me! Back to the "safety" of the Anthropocene.

Summary

What have we discovered? There are a lot of amazing fossiliferous rocks in the Bighorn Basin of Wyoming, and an especially important interval within those rocks represents the latter part of the Paleocene and the beginning of the Eocene, right around 56 Ma. In the late Paleocene, Earth was hot, and the Bighorn Basin area was quite humid. It was forested with plants that would have looked familiar to us, and with animals that likely wouldn't. Just a few thousands of years later—a geological instant, really—conditions were very different. It had become hotter still, much hotter, and quite dry. Over 90% of the plant species that had been there were gone, and the new forest, made up of plants that had earlier only grown much farther south, was likely quite sparse. The animals were also quite different, and some of the ones that were new here represent the very beginnings of some important modern groups: the artiodactyls (like deer), the perissodactyls (like horses), and the primates (like us).

In the next several chapters, we will review the conditions at the end of the Paleocene, the geological evidence for a rapid temperature spike and atmospheric change at the start of the Eocene (PETM), and how that affected the oceans and ocean life. We'll go further into how it affected the land and life on land, how long it lasted, and what finally brought it to an end.

In part 2, we will investigate the likelihood that something similar could happen in the near future as a result of anthropogenic climate change, and what that might look like for us and for the rest of life on Earth. Finally, we will explore some of the things that we can do to reduce the possibility that there will be a PETM-like runaway climate event in our future.

Chapter 2

The Late Paleocene World

This dance of plants and animals with climate change happened over vast landscapes, with forests moving from the Gulf Coast to the Rocky Mountains in just a few thousand years.

—Jonathan Bloch[1]

A S WE'VE SEEN, the configuration of the continents at the end of the Paleocene (56 Ma) was generally similar to that of today. The major differences were that the Atlantic Ocean was much narrower than it is now and the Pacific much wider, India was not yet connected to Asia, while South and North America were still separated by ocean water (the Central American Seaway), and the sea channel between South America and Antarctica (today's Drake Passage) was narrow and shallow. There were no glacial ice caps, although there could have been small glaciers on some higher mountains away from the tropics. Because of the small volume of ice, sea level was likely at least 70 meters higher than it is now, so large areas that are now low-lying land were submerged (e.g., much of Florida, large parts of Europe, the Amazon basin, Southeast Asia) and the continental shelves were quite wide.

Temperature

The Earth's overall average annual temperature during the Paleocene was in the order of 25°C, which makes it about 10°C warmer than today's average, but the differences between now and then were not consistent latitudinally (figure 2.1). In tropical regions (30° S to 30° N), the temperature was approximately 5°C warmer than now; in temperate regions, around 5° to 15°C warmer; and in polar regions, 15° to 40°C warmer.

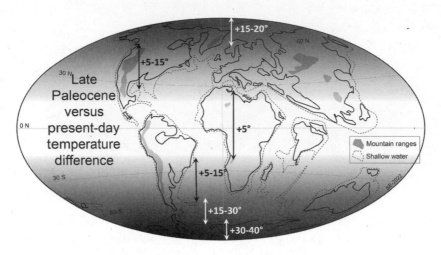

Figure 2.1: *Approximate differences in mean annual surface air temperature (°C) between the late Paleocene and today.[2]*

The difference was greatest in the southern polar region because that area, with its near complete ice cover, is currently much colder than the north polar region.

In general, therefore, the Earth had a much more evenly distributed climate during the Paleocene than it does now; there was less temperature difference between equatorial and polar regions. The average difference between the tropics and the poles was about 30°C, whereas today it is over 60°C. It is likely that few areas had an average annual temperature below 0°C during the Paleocene, although it might have been cold enough for permafrost to exist in some areas and there would have been snow.

A graphic illustration of this more evenly distributed climate can be observed on Axel Heiberg Island in the high Arctic of Canada. The Eocene rocks in that area are host to beautifully preserved fossils of hemlock, spruce, pine, redwood, gingko, birch, alder, oak, sycamore, and ferns (figure 2.2). Continental drift cannot explain this anomaly because at that time Axel Heiberg was at roughly the same latitude (83° N) as it is now. Obviously, the climate was warm enough for these plants to grow, but because of the far northern location, it was dark for six months of every year and daylight for the other six, so they had to

Figure 2.2: *Mummified stump of an Eocene Metasequoia (Dawn Redwood) on Axel Heiberg Island, Nunavut, Canada (Ansgar Walk).*[3]

grow under crazy conditions. Today Axel Heiberg Island is far beyond the tree line. The nearest trees are about 1,200 kilometers to the south; most of those are spindly and no taller than a person.

Ocean Currents

Ocean currents are critical to the redistribution of energy and matter on the Earth, and because of the differences in continental configuration, late Paleocene current patterns would have been unlike those of today. Under today's conditions, with a relatively wide Atlantic Ocean, warm salty Gulf Stream water flows north from the tropical Atlantic and Caribbean and then past Europe and Iceland into the far North Atlantic. There, the water cools down, and because of its high salinity, it becomes dense enough to sink and become what's referred to as "deep water" that flows south through the Atlantic at depths of several thousand meters and then into the Indian and Pacific Oceans, where it eventually comes back to surface. This deep flow of water that had been at surface is part of the thermohaline circulation system. Its movement is controlled by the temperature ("thermo") and salinity ("haline") of the water.[4] Formation of deep water is critical to the entire ocean current system and therefore to redistribution of heat energy and matter (dissolved nutrients) on Earth, and it has significant implications for the climate. It is very unlikely that deep water was able to form in the North Atlantic during the Paleocene because that seaway was still narrow and shallow. Instead, the Gulf Stream water likely circulated back to the south at surface. It is probable that deep water formed in other locations, and the most likely candidate is the Southern Ocean, although some may also have formed in the northern Pacific, as shown on figure 2.3.

Some other differences in ocean circulation between the Paleocene and now are that water was able to flow between the Atlantic and the Pacific through the Central American Seaway (because the Isthmus of Panama did not yet exist), water from the Tethys Sea (between Eurasia and Africa) was able to flow to and from the Pacific, and there was

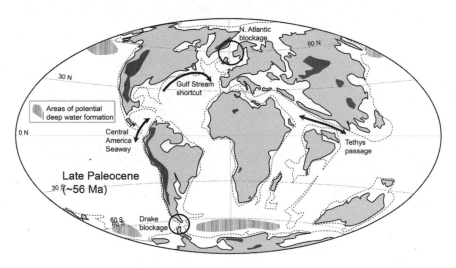

Figure 2.3: *Sites of some of the differences in ocean current patterns between now and the late Paleocene. The shaded areas are where deep water may have formed at surface, and then flowed at depth to other regions.*

no strong current around Antarctica, as there is today, because South America and Antarctica were still nearly connected at what is now the Drake Passage.

Climate Controls During the Paleocene

The Paleocene climate was controlled by the same general factors that control our climate today: atmospheric greenhouse gases (GHGs), albedo, ocean currents, orbital cycles, volcanism, and weathering, and as you would expect, some of these were quite different compared with now.

Although there are reliable ways to measure past temperatures (as we will discuss later), measuring past atmospheric gas compositions is difficult. We know air compositions accurately for the past 800,000 years from air bubbles trapped in glacial ice, but there is no glacial ice on Earth that is older than about one million years, and certainly none as old as the Paleocene (which didn't have significant glaciers). Other ways of estimating atmospheric gas levels are known as proxies, and they are based on our understanding of how various biological or mineral-forming processes respond to the proportions of different gases. We can estimate CO_2 levels in several different ways: from carbon isotope analysis of foraminifera (marine planktonic organisms), or enamel in the teeth of mammals, or sodium bicarbonate deposited in ancient lakes. We can also use boron isotope analysis of foraminifera, or the density of stomata in fossil leaves. Unfortunately, the various tools available to estimate Paleocene CO_2 levels give quite widely differing results, from around 400 to over 800 parts per million (ppm). These numbers can be compared with pre-industrial levels of 280 ppm and the current (2023) level of about 420 ppm CO_2. We know even less about the concentration of the important GHG methane (CH_4) during the Paleocene.

Although these uncertainties make it difficult to understand the extent to which GHGs were responsible for the warm climate of the Paleocene and Eocene, we do know that GHG variations could have contributed to short-term climate fluctuations, on the scale of thousands to millions of years. The important processes in that regard would have been biological: taking CO_2 out of the atmosphere and fixing it as carbon in leaves, stems, and wood; and geological: burying that organic carbon in sediments, fixing atmospheric CO_2 through weathering of rocks (i.e., converting CO_2 gas into carbonate ions dissolved in water), and emission of CO_2 through volcanism. As we'll see later, there are several other important long-term carbon storage banks on Earth—such as the carbon in permafrost—that can be released by feedbacks resulting from climate warming.

Albedo is a measure of the reflectivity of a surface (figure 2.4). Ice and snow have a high albedo, meaning that most sunlight that strikes these surfaces is reflected into space and does little to warm the planet. Bare rock and sand have much lower albedos, and vegetated surfaces are generally lower still. Open water of lakes or the ocean has the lowest albedo. Most of the sunlight that shines on open water is absorbed and converted into heat.

There was virtually no glacial ice and relatively little snow on Earth during the Paleocene, and that made the average albedo lower than it is today. Another albedo variable is the area of continents versus ocean. Although the total area of continental crust has not changed significantly since then, the high sea levels of the Paleocene made for more ocean area than there is now, and so a still lower average albedo. The latitudinal distribution of land masses is also an important albedo factor. Land is more reflective than ocean water, and because albedo is more significant at low latitudes where the sun is most intense, a concentration of continents near to the equator will result in a higher effective albedo. The continents have certainly moved over the past 56 million years, some by well over 1,000 kilometers, but since much of

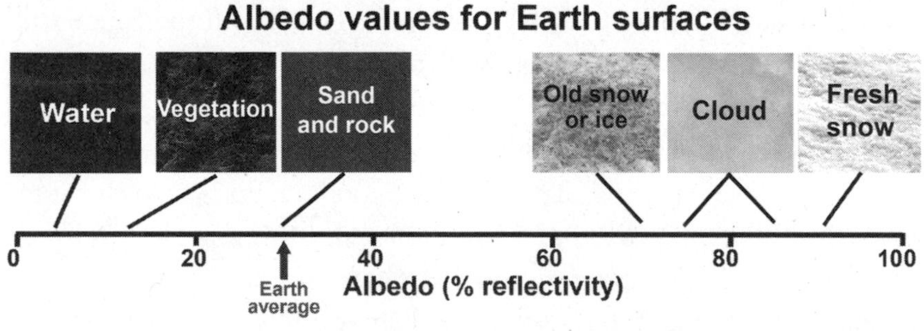

Albedo values for Earth surfaces

Figure 2.4: *Albedo values for different types of Earth surfaces*

that movement has been easterly or westerly, or because movements away from the equator (e.g., Asia and Africa) have largely been offset by movements towards the equator (e.g., North America, Australia), the movement of continents has not contributed to a big difference in albedo.

Based on all these albedo factors, it is likely that the Earth's average albedo was a few per cent higher in the Paleocene than it is now.[5] Although that sounds like very little, albedo is an important climate factor, and it is enough to account for much of the difference in temperature between then and now.

While we do know that ocean currents were significant in moving heat around on Earth during the Paleocene, we cannot be certain about which way they flowed through passages or where deep water was formed and where it might have emerged. The compelling evidence for a relatively small contrast between polar and equatorial temperatures (as illustrated in figure 2.1) argues for a robust global current system that was effective in redistributing the Earth's equatorial heat. Perhaps more important than the overall intensity of the currents is the possibility that they could change on time scales of hundreds of years. Such changes could have played important roles in the onset of the Paleocene Eocene Thermal Maximum (PETM).

Variations in the Earth's orbital parameters (Milankovitch cycles) have implications for the climate because they affect the intensity of the solar radiation (insolation) at different latitudes and at different times of the year. The elements of those variations are summarized in figure 2.5. The 100,000-year eccentricity cycle is the degree to which the Earth's orbit around the sun is elliptical as opposed to circular. The more elliptical the orbit, the greater degree to which the sun is eccentric—not centred—within that orbit. The eccentricity shown in figure 2.5 is exaggerated. The 40,000-year tilt cycle is the degree to which the axis of Earth's rotation is tilted relative to the plane of the orbit around the sun. It varies from 22.1° to 24.5° and is currently at 23.5°. The 26,000-year precession cycle is the variation in which way the Earth rotation axis points. At present, the north polar axis points to the star Polaris. At the other extreme, it points to Vega.

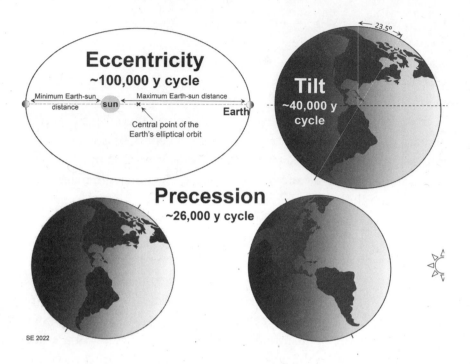

Figure 2.5: *Illustration of the Earth's orbital parameters (Milankovitch elements).*

The Milankovitch cycles are complicated, but fortunately, we don't need to understand them in detail. The key point is that, at these periods of approximately 26,000, 40,000, and 100,000 years, the intensity of insolation at various latitudes is variable. The cycles have driven the glaciation cycles over the past few million years because they determine how much insolation is received where glaciers form and grow, which is primarily at 65° N (Alaska for example).[6] Lower summertime insolation at that latitude mean less melting, and that contributes to greater accumulation of snow and so to growth of glaciers. Glacial cycles are a special case because glaciation brings its own powerful feedback mechanisms that magnify the changes in insolation of the orbital cycles. For example, as ice sheets grow, albedo increases, so it cools more, and ice sheets grow faster. And cooling also leads to storage of more carbon in permafrost and in the oceans. With or without glaciation, the Milankovitch cycles also play a role in controlling the intensity of monsoon processes because the insolation at tropical latitudes affects the degree to which the land is heated more than the oceans (which is what drives monsoons).

Milankovitch cycles have existed on Earth for all of geological time, including the Paleocene. At that time, the Earth wasn't cold enough for glacial cycles so the climate changes were not pronounced, but they likely affected Paleocene monsoon cycles, and there is evidence that they could have played a role in triggering the PETM.

Volcanism has also been a climate factor throughout Earth's history because, along with ash and lava, volcanoes emit gases into the atmosphere. The volumes of three climate-relevant gases emitted in a typical "large" eruption are illustrated on figure 2.6 and are contrasted with the volumes of those same gases that are already present in the atmosphere. The 1991 Mount Pinatubo eruption is the most recent large volcanic event. During that eruption, about 600 million tonnes of water was released into the atmosphere. That was a tiny amount compared with the 16,000,000 million tonnes already there. Pinatubo released 200 million tonnes of CO_2, also tiny compared with 2,700,000 million tonnes already

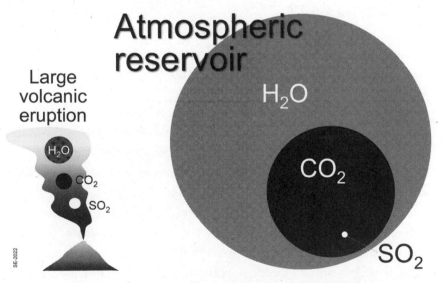

Figure 2.6: *Volumes of gases emitted during a typical volcanic eruption (left) compared with the volumes of those gases already in the atmosphere. The gas volumes are represented as spheres, not circles.*

there. On the other hand, Pinatubo released 20 million tonnes of SO_2 (sulphur dioxide), which was several times greater than the 3.2 million tonnes that was already there.

Both H_2O and CO_2 are greenhouse gases (GHGs) and so should contribute to warming, but even a large eruption like Pinatubo didn't emit enough of either to have an immediate or measurable climate impact. Volcanic H_2O doesn't remain long in the atmosphere either—just a few days—and so has almost no climate implications. On the other hand, volcanic CO_2 can remain in the atmosphere for centuries to millennia, and so, if there is a long period (thousands of years) with sustained higher than average volcanism, this extra CO_2 can make a climate-warming difference.

SO_2 gets converted into sulphate aerosols in the atmosphere (for example, tiny droplets of sulphuric acid, H_2SO_4), and those, along with fine particles of volcanic ash, block incoming sunlight, contributing to cooling. The Pinatubo eruption, which resulted in a significant increase in the atmosphere's sulphur load, contributed to spectacular sunsets and cooled the climate by 0.5° to 1°C for about two years.

As we'll see, there were some major volcanic eruptions during the Paleocene, and they may have played a role in the PETM climate change. Some took place on the seafloor along the mid-Atlantic divergent boundary, and those may have had climate impacts that were distinct from that of their gas emissions.

Weathering of rocks is a climate driver because the conversion of primary silicates (such as feldspar) to clay minerals (such as kaolinite) consumes atmospheric CO_2, leading to cooling (see the appendix, section 1). But weathering is going on all the time, and so it only affects the climate if there is a change in the rate of weathering. The global rate of weathering can be enhanced by the formation of mountain ranges, which is most commonly associated with the collision of continents due to plate tectonics. Relatively little of that happened in the late Cretaceous or the Paleocene, and that may be part of the reason for the relative warmth of those times. The Himalayas and related ranges were created by the collision of India and Asia, starting at about 50 Ma, and while that is credited as a major factor in the steady cooling of the

Earth's climate from then until now, it took place long after the climate events of interest to us here.[7]

Paleocene Land Migration Routes

Many of the land animals of the Paleocene and early Eocene had distributions across several continents. For example, the mouse-sized primate *Teilhardina* (ty-har-DEE-na)—who we met briefly in chapter 1—was present in central North America in the early Eocene, but it may have originated in southern Asia or even Europe. The rocks bearing fossils from all three of these places have dates falling within a window of about 30,000 years at the start of the Eocene. How, and why, did *Teilhardina* cover so much ground, and which way did it go? The continent distribution in the late Paleocene (figure 2.7) may provide some answers, although it is important to recognize that our understanding of Paleocene-Eocene continental positions is far from perfect.

The three known *Teilhardina* fossil locations shown as stars in figure 2.7 are all from the early Eocene rocks. A 2006 study showed that the rocks bearing *T. asiatica* in present-day China are several thousand years older than the *T. belgica* rocks (in Belgium), which are several thousand years older than the *T. brandti* and *T. americana* rocks in Wyoming.[8]

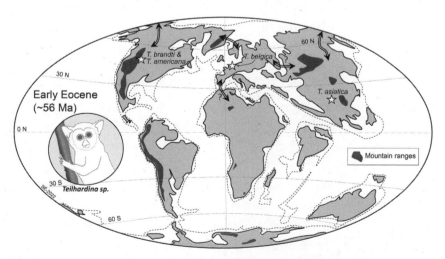

Figure 2.7: Teilhardina *localities (stars) and potential land migration routes in the early Eocene.*

A more recent study suggests that it is equally likely that the genus originated in North America.[9] Whether it started out in Asia or North America (it could not have been both), *Theilhardina* had a long way to go. Modern mammal dispersal rates are in the order of 1 to 10 kilometers per year. The distance between China and North America via Siberia and Alaska is about 14,000 kilometers, and that between China and America via Europe and Greenland is about 16,000 kilometers. If we assume the slowest of these dispersal rates, it could have taken in the order of 15,000 years for *Teilhardina* to move that far. Of course, a small tree-dwelling primate could not have trekked across deserts or even grasslands, so there would have to have been nearly continuous forest— with suitable food—along the entirety of these routes. We know that the climate would have been warm enough for forests in all these areas, but we don't know about the potential for deserts or grasslands. Apart from that, it is likely that *Teilhardina* would have found it difficult to cross a body of water even just a few kilometers wide.

If we take the continental distributions of figure 2.7 at face value, the route from Asia to North America (or vice versa) via Siberia and Alaska looks doable. The route from Asia to Europe appears to be more of a challenge, but it must have been possible. The route from Europe to North America via Greenland (or vice versa) looks quite doubtful, but as noted, it is possible that the map is not exactly correct, and that there was a land passage from Europe to Greenland at that time.[10]

But why? Why would these tiny timid creatures leave southern Asia or central North America and just keep moving? Not just because they could, but because they had to. The climate was warming significantly during this time—although not as quickly as it is now—and vegetation communities were changing as a result. Generation after generation of *Teilhardinas* became refugees. They were compelled to leave the places they were living in and move towards more suitable habitat, and the changes in climate and vegetation determined the directions they went.

Of course, as we'll see later, *Teilhardina* wasn't the only animal forced to move around at this time. Others might have followed similar or different paths, but the evidence indicates that it was possible for land animals to disperse across all parts of the northern continents: North

America, Europe, Asia. On the other hand, South America, India, Antarctica, and Australia were all isolated for land animals, and Africa may have been as well.

Summary

Some of the important features of the late Paleocene can be summarized as follows:

- The Earth was about 10°C warmer than it is today, and there was less contrast in temperature between the tropics and the poles. Forests could grow at almost any latitude.
- Because of the distribution of the continents, ocean currents were quite different from those of today. There was no deep-water formation in the far northern Atlantic, but there may have been deep-water formation in the Southern Ocean.
- Water could move between the Atlantic and the Pacific through the Central America Seaway, but there wasn't a strong current around Antarctica.
- The CO_2 level of the atmosphere was likely higher than it is today (at least higher than the pre-industrial level of 280 ppm), but there is conflicting evidence as to how high it might have been.
- With little or no glacial ice, very little snow, and high sea level, the Earth's albedo was lower than it is now (it was less reflective), low enough to account for a significant part of the difference in temperature.
- There was significant volcanism during the Paleocene, and it is possible that this contributed to climate changes.
- There is strong evidence that the late Paleocene/early Eocene continental configuration allowed for dispersion of land animals between Asia, Europe, and North America, and that many animals and plants were compelled to migrate because of the changing climate.

Chapter 3

Understanding Past Climates

> *The deep sea is the largest museum on earth, it contains more history than all the museums on land combined, and yet we're only now penetrating it.*
>
> —Robert Ballard[1]

SEDIMENTARY ROCKS and the sediments that are their precursors—both in the oceans and on land—provide an extraordinary record of Earth's history, including the history of life on Earth and the history of our changing climate. Sedimentary layers come from two main sources: particles (grains) that have been produced by the physical breakdown of rocks and dissolved ions that have been produced by the chemical breakdown of rocks. The particles include tiny fragments of clay and sand-sized fragments of other minerals like quartz and feldspar, and larger pieces of preexisting rocks. The dissolved ions include those of calcium, sodium, carbonate, chloride, silica, iron, magnesium, and many others.

All of this material is brought to the oceans by murky grey, green, brown, and even red rivers laden with clay, silt, and sand (or larger particles when in flood). The larger particles are deposited near to shore, but the tiny clay fragments stay in suspension for weeks to decades and get washed out into the open ocean, where they finally settle onto the deep seafloor. The ions of calcium, bicarbonate, iron, silica, and magnesium are also dispersed across the oceans.

These ions are amongst the important chemicals needed by marine organisms to thrive. Iron, for example, is critical for the process of photosynthesis in phytoplankton, and silica is used by organisms like microscopic diatoms and radiolarians to construct their hard parts (tests).

Calcium and bicarbonate are used by a wide range of marine organisms to build living structures and shells. Although that may make you think of clams, oysters, snails, and corals, the most important organisms, from our perspective here, are the foraminifera ("forams" for short) and coccolithophores (figure 3.1), which are two types of single-celled protists.[2] Many of them drift across surface waters, but some inhabit the middle layers and others the dark ocean depths. They have existed since the Cambrian—over 500 million years— and include over 50,000 species (both alive and extinct).

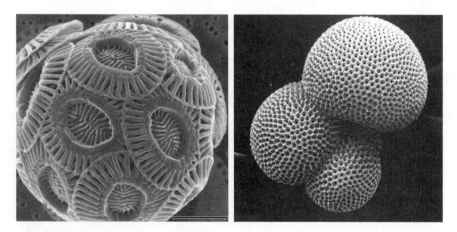

Figure 3.1: *Coccoliths constructed by an* Emiliania *coccolithophore (left) and tests of a* Globigerina *foraminifera (right). The coccoliths are very small (about 0.002 mm across) while the foram tests are much bigger, about 0.25 mm across. All are made of calcium carbonate. (Coccoliths by Alison R. Taylor.[3] Foraminifera by Hannes Grobe.[4])*

Seafloor Sediments

Most forams and coccolithophores build tests ("shells") out of the mineral calcite (also known as calcium carbonate or $CaCO_3$), and when they die, the living part wastes away while the test slowly sinks towards the seafloor. There is an important reason for the use of the word "towards" rather than "to" in the previous sentence. Physics would have the tests settle right to the bottom, but chemistry doesn't necessarily allow that. Calcite is insoluble in surface water (which is good news for organisms that depend on shells and tests), but it becomes soluble at a water depth of around 4,500 m. This is known as the carbonate compensation depth, or CCD[5]; it varies with latitude and also with temperature, pressure, and, most important, acidity.

Most parts of the oceans are deeper than 4,500 meters. Shallower depths are only common in areas within a few 100 kilometers of shore, or along the spreading ridges, like the Mid-Atlantic Ridge, or on the flanks of submarine volcanoes (figure 3.2). Foram tests, and any other calcite parts that sink through the water column, start to dissolve when they reach a depth of about 4,500 meters, and so they are converted back into calcium and carbonate ions.

The main components of sediments in the open ocean—away from shore and from the mouths of large rivers—are clay minerals (which are mostly derived from weathering on the continents) and calcite (which is mostly from marine organisms). Because of the abundance of marine life, the sediments that accumulate in the shallower areas are rich in the calcium carbonate parts of forams and coccolithophores, and so those sediments are dominated by calcite. They tend to be light coloured, even white if there is enough calcite. The chalk of the white cliffs of Dover is rich in coccoliths. Sediments that accumulate in water deeper than

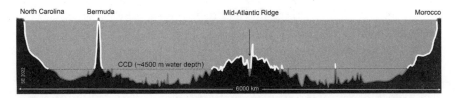

Figure 3.2: *Generalized profile of the Atlantic Ocean between North Carolina and Morocco (The carbonate compensation depth is ~4,500 meters. White denotes carbonate sediment accumulations. In all other areas, the sediments are dominated by clay. The vertical exaggeration is about 1,000 times, which means that the seafloor is not nearly as rugged as it looks here. If you were to drop down to any location on this cross-section, even in a place that appears to be steep, the floor of the sea would appear to be as flat as any plain on the continents.) The Mid-Atlantic Ridge is the divergent boundary between the North America and Africa Plates. (Based on profiles drawn in 1959 by Marie Tharp and Bruce Heezen of the Lamont Doherty Geological Observatory, Columbia University.[6])*

4,500 meters are dominated by clay. They tend to be grey or brown, but can also be reddish if there is source of oxidized iron, or even dark (almost black) if there is a significant amount of organic matter.

One way to envision this is to think of the seafloor peaks and ridges above about 4,500 meters depth as being "snowcapped," although it isn't snow of course, it is white calcite (figure 3.2). The rest of the seafloor is covered in grey or brown mud dominated by clay minerals.

Climate Data from the Oceans

OK, so why does all of this matter? It matters because parts of the ocean floors are over a hundred million years old, and they have been accumulating sediments for all that time. In most areas of the open ocean, the rate of accumulation is very slow: less than a millimeter per century, but it's also very consistent, and so hundreds of meters of sediment have accumulated over those millions of years. Those sedimentary layers represent a massive library (or "museum," in Robert Ballard's words) of information about conditions during the Earth's past, and we are getting increasingly good at reading that sedimentary record. Forams are a critical part of interpreting past temperatures and atmospheric compositions because they use carbon and oxygen from the atmosphere and the ocean to make their $CaCO_3$ tests.

For over fifty years, marine scientists from around the world have cooperated on a massive project to sample the sedimentary layers of the seafloor. This is accomplished by drilling from ships, first by lowering the drill pipe through hundreds or thousands of meters of water, and then by drilling into tens or hundreds of meters of sediment and taking core samples. Figure 3.3 shows some of the locations where seafloor coring has been done.

Cores collected from these locations are carefully studied and described. Any ash layers from large volcanic eruptions are precisely dated using isotopic techniques, and the ages of intervening layers are interpolated from that. Samples are extracted and carefully processed to separate the tests of forams and any other marine organisms, and these are sorted by genus and species. Oxygen and carbon isotope ratios are then determined using the foram subsamples.

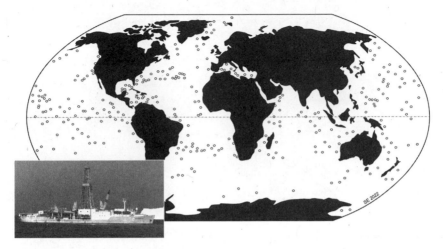

Figure 3.3: *Some of the hundreds of locations (open circles) where sediment cores have been acquired from the ocean floor. Inset: the drilling ship* JOIDES Resolution, *which was used to acquire cores at most of the sites shown on the map. (The inset of the* JOIDES Resolution *in the Weddel Sea (Antarctica) is by Hannes Grobe.[7])*

Oxygen Isotopes

The isotopic[8] proportions of oxygen and carbon are key to understanding past conditions, and so we need to know a little about those. The most abundant isotope of oxygen, which makes up 99.7% of the oxygen on Earth, has 8 protons and 8 neutrons (figure 3.4). That makes it element number 8, with an atomic mass of 16. The main other isotope of oxygen, oxygen-18, has 10 neutrons and 8 protons, so it's still element number 8, but it has an atomic mass of 18. It comprises only 0.2% of all oxygen. Both of these oxygen isotopes are stable, or non-radioactive.

While ^{18}O (shorthand for "oxygen-18") is a little more massive than ^{16}O, the two are both still "oxygen," and their chemical behaviors are *almost* exactly the same. The difference is that heavier ^{18}O is a little less nimble than ^{16}O, so when there is a reaction that results in a change in state—say when dissolved oxygen in seawater gets incorporated into the solid calcite molecules of a foram test, ^{18}O is a little more likely to want to go to the more solid form than is ^{16}O. This results in the solid form having a slightly higher ^{18}O to ^{16}O proportion than that of the oxygen in the water. This process is known as fractionation. The

Oxygen isotopes

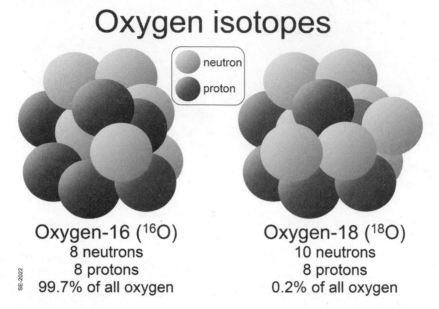

Oxygen-16 (^{16}O)   Oxygen-18 (^{18}O)
8 neutrons   10 neutrons
8 protons   8 protons
99.7% of all oxygen   0.2% of all oxygen

Figure 3.4: *Representations of the nuclei of the oxygen-16 and oxygen-18 isotopes.*

critical thing here is that this tendency for ^{18}O to get comfortable in a more solid molecule is also temperature dependent. The colder it is when the reaction takes place, the stronger the tendency for ^{18}O fractionation. In other words, if we have several different foram tests that formed at different temperatures, those formed at a lower temperature—say 10°C—will have a greater proportion of ^{18}O than those formed at a higher temperature—say 20°C.[9] So, by collecting sediment samples at different depths of the seafloor, and painstakingly picking the foram tests out of them, and then carefully analyzing those for their ^{18}O to ^{16}O proportions, we can determine past global temperature variations.

Temperature Variations in the Cenozoic

Marine scientists have acquired oxygen isotope data from tens of thousands of ocean sediment core samples and compiled them to create the record of global temperature variations for the Cenozoic (the past 66 million years) that is shown in figure 3.5.

The average global temperature was approximately 25°C in the Paleocene. It rose to nearly 30°C in the early Eocene and, since then,

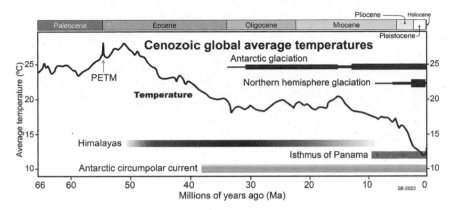

Figure 3.5: *Generalized global temperature variations for the past 66 million years, based on a compilation of oxygen isotope data from seafloor sediments from many locations. The widths of the glaciation indicators are generally proportional to the extent of glaciation at those times. (The temperature curve is based on oxygen-18 isotope analysis of foraminiferal tests extracted from multiple marine sediment cores and is originally from Zachos, J., et al.[10] Glacial intensity indicators are from Westerhold et al.[11])*

has systematically dropped by about 14°C. Much of that decline is attributed to the strong weathering related to the growth and then the rapid erosion of the Himalayan Range. That process consumes atmospheric CO_2. (See the appendix, section 1.)

Other events have also had an impact on global temperatures over this time period. For example, the formation of the Antarctic Circumpolar Current isolated Antarctica from the rest of the oceans and likely contributed to the start of glaciation there at around 35 Ma. The closure of the Caribbean Seaway by the Isthmus of Panama at around 10 Ma changed the way currents flow in the central and northern Atlantic, and may have been responsible for the cooling trend that led to northern hemisphere glaciation starting around 4 Ma.

A prominent feature in figure 3.5 is the sharp spike in temperature at the Paleocene-Eocene boundary. This is the Paleocene-Eocene Thermal Maximum (PETM), and it's why we're here. The PETM temperature anomaly was first detected in marine sediments in 1991.[12] Although it appears, in that diagram, to be an almost instantaneous spike and then an equally fast drop in temperature, when the data are plotted on a shorter

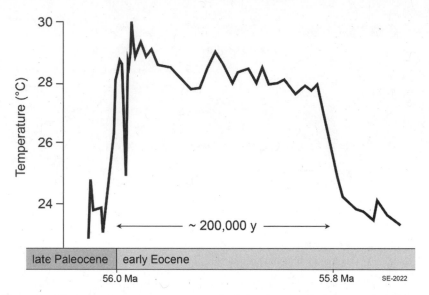

Figure 3.6: *Variations in temperature across the Paleocene-Eocene transition based on oxygen isotopes from foraminifera in marine sediments from a location near to the eastern coast of the United States.*[13]

time scale, it is evident that the warming took place over a few thousand years and the high temperatures persisted for close to 200,000 years (figure 3.6).

Carbon Isotopes

Carbon also has isotopes (in fact all elements do), and the most common ones are carbon-12 (aka ^{12}C, 99% of all carbon), carbon-13 (^{13}C, 1%), and carbon-14 (^{14}C, ~0.1%) (figure 3.7). The ^{12}C and ^{13}C isotopes are stable (they last forever), while ^{14}C is unstable (radioactive), with a half-life of 5,730 years. That's why ^{14}C is the basis for carbon dating, and also why it isn't of any interest here.

Like oxygen isotopes, the ^{12}C and ^{13}C isotopes behave a little differently from each other and are subject to fractionation. In general, ^{12}C is favored over ^{13}C when carbon gets fixed into biological materials, although the degree to which that happens varies with the biological process. As illustrated in figure 3.8, CO_2 in the ocean and the carbon structures formed by marine animals in the ocean have ^{13}C to ^{12}C

Carbon isotopes

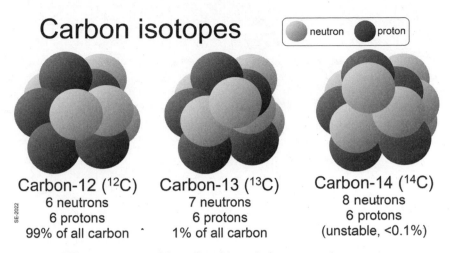

| | neutron | | proton |

Carbon-12 (^{12}C)
6 neutrons
6 protons
99% of all carbon

Carbon-13 (^{13}C)
7 neutrons
6 protons
1% of all carbon

Carbon-14 (^{14}C)
8 neutrons
6 protons
(unstable, <0.1%)

SE-2022

Figure 3.7: *Representations of the nuclei of the carbon-12, -13, and -14 isotopes*

ratio differences close to 0, while CO_2 in the atmosphere has a slightly negative difference. Plants, soil, CO_2 within soil, and fossil fuels have moderate negative ^{13}C differences, and the methane reservoirs have the strongly negative ^{13}C differences. That includes permafrost methane, and thermogenic methane (which is produced when volcanic magma comes into contact with seafloor sediments). The reservoir with lowest ^{13}C difference is methane hydrate, which is an ice-like combination of water and methane that mostly exists within sediments on the seafloor.

Variations in the ^{13}C/^{12}C differences of various C reservoirs

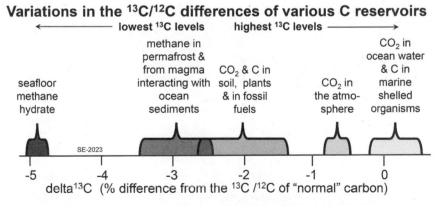

← lowest ^{13}C levels   highest ^{13}C levels →

| seafloor methane hydrate | methane in permafrost & from magma interacting with ocean sediments | CO_2 & C in soil, plants & in fossil fuels | CO_2 in the atmo-sphere | CO_2 in ocean water & C in marine shelled organisms |

SE-2023

delta^{13}C (% difference from the ^{13}C /^{12}C of "normal" carbon)

-5   -4   -3   -2   -1   0

Figure 3.8: *Typical ranges of ^{13}C to ^{12}C proportions in different Earth reservoirs.*[14]

Carbon in rocks (e.g., limestone and fossil fuels), and in soil, permafrost, and methane hydrate is very stable; it can remain fixed for thousands to millions of years, and so can be thought of as long-term carbon storage. On the other hand, most of the carbon in the air, water, and plants is exchanged relatively rapidly (days, weeks, years, centuries). Of these "rapid-exchange" carbon reservoirs, the ocean is by far the largest, with about 60 times as much carbon as in the atmosphere and about 20 times as much as in living plants. The ^{13}C ratio of ocean water determines the ^{13}C ratio of the shells and tests that are made by forams and other marine organisms, and so variations in the carbon isotope ratios of forams provide us with a record of past carbon isotope changes in the ocean and, indirectly, the atmosphere.

The ^{13}C to ^{12}C ratio (i.e., the δ^{13}C value) of ocean water CO_2 tends to stay quite constant over geological time (close to 0% for the deep ocean and a little higher for the shallow ocean), but from time to time in the past, that δ^{13}C value has changed, and of course—no prize for guessing—one such change took place during the PETM.

The ^{13}C differences for the same set of marine sediment samples that were used to estimate the temperature variations shown in figure 3.6 are portrayed in figure 3.9. Prior to the PETM temperature rise, the ^{13}C differences were close to 0%. As the temperature rose, the

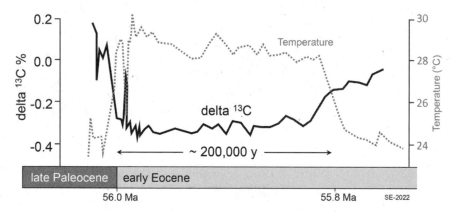

Figure 3.9: *Variations in ^{13}C/^{12}C ratios (heavy line) across the Paleocene-Eocene transition based on foraminifera in marine sediments from near to the eastern coast of the United States. Dotted line is temperature, as shown in figure 3.6.*

[13]C differences at this location dropped to almost -0.4% and stayed low for about 200,000 years. This change is called a "carbon isotope excursion" or CIE. Although we don't exactly know what caused it, the likely suspects are those listed towards the left-hand side of figure 3.8. Remembering that some of those reservoirs—for example, permafrost methane, thermogenic methane, or methane hydrate—have very low ^{13}C levels, we can see that the release of a significant amount of that carbon into the ocean could change the overall ocean ^{13}C ratio enough to account for a CIE like the one observed in these sediments.

The percentage of calcium carbonate ($CaCO_3$) is another important variable that is typically measured in seafloor sediment samples. That information is useful because it provides a measure of variations in the solubility depth of $CaCO_3$. When atmospheric CO_2 levels increase, the acidity of the oceans increases,[15] and the depth at which $CaCO_3$ becomes soluble (the CCD) decreases, it shallows. This is illustrated in figure 3.10 for a location off the west coast of southern Africa. During the late Paleocene, this seafloor site was above the CCD (it was on the flank of one of the snowcapped peaks), so calcite was able to accumulate in the sediments. In the early Eocene, the CO_2 level in the atmosphere went up, the oceans became more acidic, and the CCD rose quickly, such

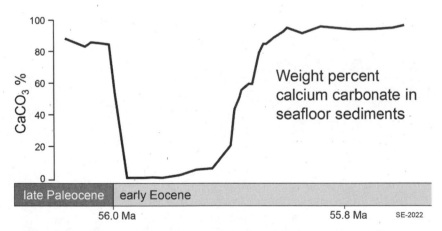

Figure 3.10: *Variations in weight % calcium carbonate ($CaCO_3$) in core from Ocean Drilling Program hole 1267B off the west coast of southern Africa. This profile represents the same time interval as that shown in figures 3.6 and 3.9.[16]*

that calcite shells and fragments raining down from above dissolved before they reached the seafloor at this site, and the sediment that accumulated during that time had almost no calcite.

Box 3.1 Climate Models

We have examined some of the types of ocean records that can be used to investigate past climate and some of the methods—such as isotopic analyses—that allow us to understand what went on in the past. These indirect, or "proxy" indicators, do not provide a first-hand reading of ancient climates, as we might get if we could travel back in time with a thermometer or a barometer, but they work reasonably well because we understand the physics and chemistry of the proxies. For example, we know that the oxygen isotope composition of a foram is proportional to the temperature of the water in which it lived.

Another way to investigate past (and present, and future) climates is through climate models, also known as general circulation models (GCMs). These are numerical models of the Earth's climate, or of some part of it (or even of another planet), that can be used to calculate the rates of transfer of energy and matter in the atmosphere and oceans. Understanding these rates allows us to estimate how the climates of those locations will change over time if we change the input parameters. For example, we could gradually change the greenhouse gas content of the atmosphere in the model to see how temperature, humidity, wind, rain, and ocean currents will change. Or we could change the distribution of incoming solar energy (to simulate Milankovitch cycles) to see how that affects the growth or decay of glacial ice, or other aspects of the climate.

How a GCM Works

A GCM is based on a cellular representation of the Earth, as is illustrated here. The Earth's surface is divided into cells in a horizontal sense, and then the atmosphere, oceans, ice, vegetation, soil, and solid rock of the crust are divided into cells in a vertical sense. The GCM is initialized with information on the parameters of every cell, such as: the temperature and GHG (greenhouse gas) composition of the atmosphere; the temperature, salinity, and other chemical features of seawater; the type of vegetation and various others. It also considers the incoming radiation from the sun and the outgoing infrared radiation from the Earth, and the effect that each cell has on its neighboring cells, both horizontally and vertically. Using a set of complex equations based on well-understood geological, physical, chemical, and biological processes, the GCM starts calculating how the different cells interact with each other, and how those interactions change over time, and with changes in parameters like GHG levels.

For example, depending on its density, a parcel of air may rise, and that will produce wind. Depending on the humidity, clouds may form, and depending on the temperature, it may rain or snow. Ocean currents will be impacted by wind and by the differences in density of water cells. Ice may form on land or the ocean, and that will affect albedo and so the

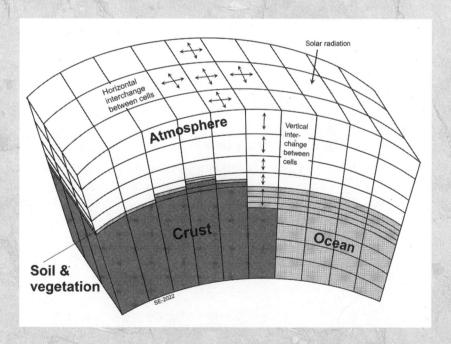

degree to which solar energy is effective in heating the Earth. Vegetation will grow to varying degrees, and that will affect the composition of the atmosphere and also the albedo.

Evolution of GCMs

The GCM technique was first used in the late 1950s. At that time, only a restricted range of parameters could be modeled, and the resolution was very coarse. In the 1970s, only solar radiation, atmospheric CO_2, and rain were included. In the 1980s, surface albedo and clouds were added. In the 1990s, volcanism, sulphate aerosols, and ocean currents could also be modeled. In the early 2000s, the carbon cycle, aerosols (e.g., dust and sea spray), rivers, and deep-ocean circulation were incorporated, and by 2010, atmospheric chemical reactions and vegetation were included.

Early GCMs suffered from poor resolution because of computational limitations. As illustrated in the figure on the right, most whole-Earth models had cells in the order of 500 kilometers in the early 1990s. That improved to about 250 kilometers in 1995, to 180 kilometers in 2001, and to around 110 kilometers in 2007. Some whole-Earth models used today have resolutions in the order of 50 kilometers.

But more important than spatial resolution is that the skill of GCMs is also improving. In its 2021 report, the IPCC provides a comparison of the accuracy of GCMs over a 17-year period from 2004 to 2021.[17] Near-surface air temperature determinations of models in use between 2004 and 2008 correlated with actual observation at a level of 98.5%. Models used between 2018 and 2021 showed average correlations of 99.2%. The accuracy of precipitation modelling is lower but has also improved significantly over time.

Example of improvement in resolution of models used in the Intergovernmental Panel on Climate Change (IPCC) reports, from 1990 to 2007. (This image is from AR4 Climate Change 2007.[18]*) The figures show how successive generations of GCMs increasingly resolved northern Europe. These illustrations are representative of the most detailed horizontal resolution used for short-term climate simulations. The century-long simulations cited in IPCC Assessment Reports after the* First Assessment Report *(FAR) were typically run with the previous generation's resolution. Vertical resolution in both atmosphere and ocean models is not shown, but it has increased comparably with the horizontal resolution, beginning typically with a single-layer slab ocean and 10 atmospheric layers in the FAR and progressing to about 30 levels in both atmosphere and ocean. Both the horizontal and vertical spatial resolutions of GCMs are increasing as a result of greater computing power and efficiencies in coding.*

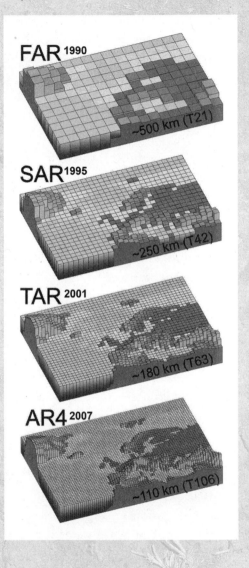

FAR 1990 ~500 km (T21)

SAR 1995 ~250 km (T42)

TAR 2001 ~180 km (T63)

AR4 2007 ~110 km (T106)

Summary

The ocean floor is covered with layers of sediments that are made of materials derived from weathering of rocks on land—the key components being clay minerals and dissolved ions such as calcium (Ca^{2+}) and bicarbonate (HCO_3^-). These ions are used by marine organisms to make their hard parts, and the most important of these is the mineral calcite ($CaCO_3$).

The oxygen and carbon in calcite have isotopic proportions that provide useful information about the environmental conditions that existed while the organism was alive, and we can access that information through chemical and isotopic analysis of sediment core samples collected from the seafloor.

Cores from many different locations have been compiled to show variations in global temperatures over the past 66 million years. Although there has been a general and significant cooling trend over that time, there are some pronounced anomalies, a key one being the Paleocene-Eocene Thermal Maximum (PETM) at 56.0 Ma.

Oxygen isotope data shows that ocean temperatures rose by around 6°C at the start of the PETM, and stayed high for almost 200,000 years. The period of extreme warming coincides with a significant drop in the ^{13}C content of seafloor calcite (a carbon isotope excursion, or CIE), indicating that the temperature change was associated with the release of carbon from a long-term storage reservoir that had very low ^{13}C levels.

At many moderately deep seafloor locations, the CIE coincides with a sharp reduction in the calcite content of marine sediments, implying that carbonate shell fragments were dissolved because of a shallowing of the carbonate compensation depth (CCD) caused by an increase in the acidity of ocean water.

Chapter 4

Changes in the Oceans
During the PETM

With every drop of water you drink, every breath you take, you're connected to the sea. No matter where on Earth you live. Most of the oxygen in the atmosphere is generated by the sea.

—Sylvia Earle[1]

IT GOES WITHOUT SAYING that the oceans are a critical part of the Earth's climate system, so if we want to understand how the climate changed during the PETM, we need to understand how the oceans changed. The important differences included increases in the temperature and acidity of ocean water, decrease in the dissolved oxygen level, rearrangement of ocean currents, and rise in sea level. These various changes dramatically affected life in the oceans, and on land.

Ocean Temperature

The PETM temperature and carbon isotope anomaly was first recognized in 1991 in data from foram-bearing ocean sediment samples from a site in the Southern Ocean.[2] Its existence has since been confirmed in seafloor sediment samples from around the world, and in sediments formed on land. (Those are discussed in chapter 5).

Figure 4.1 shows global differences in near-surface ocean water temperatures between the last several thousand years of the Paleocene and the main part of the PETM, based on data from several ocean-drilling sites and a global circulation model (GCM). Ocean temperatures increased by more than 6°C over large areas of the Southern Ocean, the northern Atlantic Ocean, the northern Tethys Sea (between Africa and Europe), and parts of the northern Pacific Ocean. Most of the

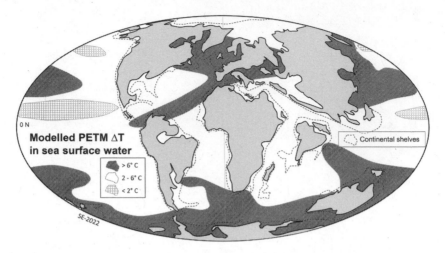

Figure 4.1: *Changes in sea surface water temperature (°C) in the PETM (early Eocene) compared with the late Paleocene based on a General Circulation Model.[3]*

remaining ocean areas experienced PETM temperature differences in the plus 2° to 6°C range. In general, therefore, the model shows that temperate regions experienced greater PETM ocean-water temperature increases than tropical regions. This is generally consistent with the current trend of greater anthropogenic climate change in temperate and polar regions.

Ocean Currents

While it is difficult for us to reconstruct the patterns of Paleocene or Eocene surface ocean currents, we can make some inferences about the nature of subsurface current flows based on detailed analysis of carbon isotopes and other parameters in ocean sediments. Recall from chapter 2 that, under present-day conditions, warm salty water from the tropical Atlantic flows north (via the Gulf Stream) and gradually cools as it does so. In the area north of Iceland, this cold and salty water becomes the densest water anywhere in the oceans, and it sinks, forming deep water that flows south at depth as part of the thermohaline circulation system, and only comes back to surface in the Indian and Pacific Oceans. That could not happen during the Paleocene because the North Atlantic was too narrow and too shallow. Instead,

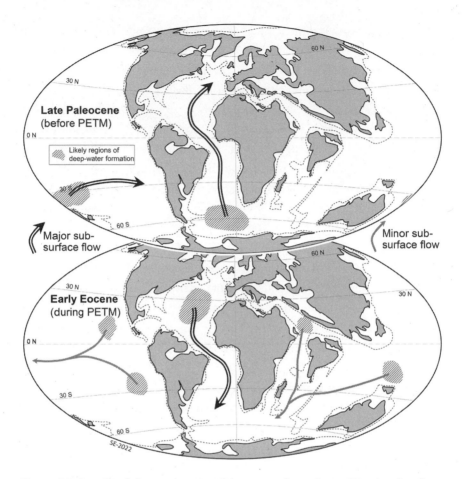

Figure 4.2: *Postulated changes in areas of deep-water formation and in subsurface flow patterns between the late Paleocene (top) and the early Eocene (bottom). Dense water sinks within the shaded areas and then flows in the general direction of the arrows.*[4]

carbon isotope records indicate that deep-water formation took place in the Southern Ocean and the southern parts of the Pacific Ocean (figure 4.2).

The high temperatures of the PETM led to an increase in the intensity of the hydrological cycle. There was more evaporation (especially in the tropics), the warmer atmosphere was able to hold more moisture, and there was more rain. Although rainfall patterns were quite spatially variable (as they are today), this had the overall effect of transferring fresh water, as rain, from tropical regions into temperate regions. The temperate and polar parts of the ocean surface waters were thus diluted, and the tropical parts became saltier. This change appears to have contributed to a pronounced shift in the locations of deep-water formation from the far south in the late Paleocene, to the tropics and subtropics early in the PETM (figure 4.2). In the late Paleocene, the formation of deep water was likely mostly driven by a combination of the low temperature of the near-polar waters. During the PETM, the driving factor for deep-water formation was most likely the high salinity of the warm waters in tropical regions. This allowed relatively warm water to sink to depth.

Apart from changes in the location of deep-water formation and in the directions of deep currents, there is evidence of reduction of the magnitude of very deep flows during the PETM. A GCM study has indicated that while there was strong circulation to depths of over 4,000 m in the pre-PETM ocean, during the PETM strong circulation was likely to have been limited to a depth of about 1,500 meters, with effective stagnation below that depth.[5]

As we'll see, these profound changes in the patterns of deep-ocean circulation had significant implications for ocean life during the PETM, and for the potential for strong climate feedbacks.

Sea-level Rise

Figure 4.3 shows a number of sites where there is evidence for sea-level rise during the PETM. The widespread extent of these sites implies global sea-level rise due to an increase in the volume of the oceans, but the magnitude of rise is not well understood. Some studies have implied sea-level rise in the order of 20 to 30 meters,[6] while others are consistent with a number closer to 5 meters.[7] The difference in these estimates is important because, while a sea-level rise of up to 5 meters could have been caused simply by the warming of ocean water, and its consequent expansion, any amount more than that would have required melting of a significant amount of glacial ice or permafrost. There was very little (if any) glacial ice at the start of the PETM (and not enough to produce 20 meters of sea-level rise), but there may have been enough

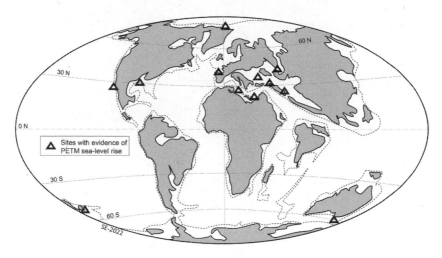

Figure 4.3: *Locations with evidence for a rise in sea level in the early PETM.*

permafrost in Antarctica, and in the northern parts of North America and Asia to lead to sea-level rise of more than 5 meters.

Ocean Water Anoxia

Anoxia (depletion of oxygen) in shallow coastal waters can be caused by a number of factors, including warming (warm water can hold less oxygen than cold water), introduction of large amounts of organic matter from the land that then gets broken down and oxidized (consuming dissolved oxygen), and introduction of nutrients, such as nitrate or phosphate, that stimulate the growth of algae, which also gets broken down and then oxidized. In deep waters, the main cause of anoxia is a reduction of deep circulation because water that was on the open-ocean surface is well oxygenated, but if it is no longer reaching a significant depth, it cannot oxygenate the deep ocean.

Ocean water oxygen is critical to the existence of many types of organisms, and so it is important to understand the incidence of anoxia during the PETM. As shown in figure 4.4, evidence of anoxia is widespread in PETM oceanic sediments in both shallow and deep settings.

Modelling studies support the evidence for deep-ocean anoxia and imply that it was likely to have been stronger and more extensive in the Atlantic basin than in the Pacific.[8] That said, it is unlikely that severe

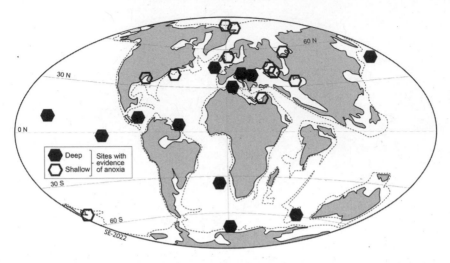

Figure 4.4: *Locations with evidence for anoxia (low oxygen levels) in the early PETM.*

anoxia (known as euxinia) existed in the deep ocean because PETM sediment samples from many sites have red coloration (indicative of iron-oxide minerals) and no evidence of hydrogen sulphide (which would indicate euxinia).

PETM deep-water anoxia is likely to have been caused by the reduction in overall circulation intensity and particularly in deep circulation, which is described above. Shallow-water anoxia may be a product of higher temperatures and an enhanced hydrological cycle (more rain) that brought more nutrients and more organic matter from the land into the oceans.

Ocean Acidification

The pH (acid-base balance) of ocean water is critical to the existence of organisms that use calcite ($CaCO_3$) to make hard parts, such as shells, tests, and coral structures. That is because calcite structures are increasingly difficult to make and maintain if the water becomes acidic. This is becoming a problem in the modern ocean due to anthropogenic climate change, and it was a factor during the PETM.

Compelling evidence for PETM ocean acidification is illustrated in figure 4.5. The data are from three Ocean Drilling Project cores acquired from the Walvis Ridge area off the west coast of southern Africa. The cores were intentionally collected at different water depths, and the depths that pertained during the Paleocene are shown. In the late Paleocene and after the PETM, all these locations were above the level of the carbonate compensation depth (CCD)[9] (which was likely around 4,000 m at the time), and so calcium carbonate did not dissolve at these depths and the sediments that accumulated are rich in calcium carbonate, in the 80% to 90% range.

At the start of the PETM, the calcium carbonate level at the deepest site (ODP-1262) dropped precipitously and remained close to 0 for about 100,000 years. At the middle-depth site (ODP-1266), carbonate levels also dropped quickly at the start but began to recover after about 50,000 years. At the shallowest site (ODP-1263), the proportion of carbonate dropped, but not quite to 0%, and started to recover after only

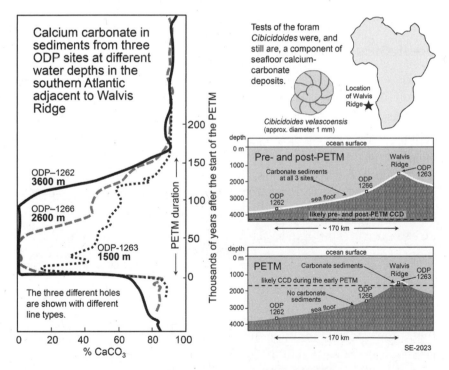

Figure 4.5: *PETM variations in CaCO₃ content of sediments at different depths in the southern Atlantic (CCD is the depth below which CaCO₃ is soluble).*[10]

a few thousand years. From this it can be interpreted that the CCD in the southern Atlantic shallowed quickly by around 2,000 m in the early PETM, and since the acidity of the water is a primary CCD controlling factor, we can conclude that the oceans were significantly acidified during the PETM.

It is estimated, based on proxy data, that pH actually decreased between 0.3 to 0.4 units during the PETM.[11] In other words, the oceans became more acidic, and considering their huge volume, a change of that magnitude is phenomenal. Modelling analysis suggests a drop in pH of greater than 0.3 units in the North Atlantic Ocean and Tethys Sea, a drop of 0.25 to 0.3 in most other regions of the ocean, and a drop of less than 0.25 in tropical oceans.[12]

By comparison, the anthropogenic increase in the atmospheric CO_2 level has, so far, resulted in a pH decline of about 0.1 units in the surface waters of the North Atlantic and the Southern Ocean, and a decline of 0.06 to 0.08 units in most other regions.

Ocean Carbon Isotope Excursion

The ^{13}C proportions of foram tests in ocean sediment cores dropped sharply in the early part of the PETM, remained low for most of the event, and recovered slowly towards the end (see figure 3.10). The carbon isotope excursion (CIE)—or change in carbon isotope levels—was between -0.1 and -0.3% delta ^{13}C, typically on the low side of that range in deep-water sites[13] and higher in shallower sites.[14] This change is attributed to a change in the ^{13}C level of ocean water, and that is likely to have resulted from the release to the atmosphere of massive amounts of carbon with low ^{13}C levels from reservoirs such as permafrost or deep-ocean sediments. This is described in more detail in chapter 6.

Marine Extinctions and Extirpations

What was living in the Paleocene ocean? Almost everything that lives there now was in the water then, except for the marine mammals—whales, seals, etc.—that didn't evolve (from land mammals) for another several million years. And then, as now, tiny organisms made up most of the ocean's biomass. Paleocene protists (which includes algae and forams), bacteria, and tiny arthropods comprised many times the biomass of all of the larger organisms, such as fish, molluscs, larger arthropods, corals, and echinoderms put together. And since the protists, bacteria, and some of the arthropods are all very small, they outnumbered the larger organisms by a huge factor. This means that it is a lot easier to understand what happened to tiny organisms during the PETM than what happened to larger ones, and so that's what we'll focus on here.

The most significant biological change of the PETM was the extinction of between 35% and 60% of all deep-water forams. This is important because large-scale deep-sea extinctions are extremely rare in the geological record. Changes in the vast deep ocean are typically very slow (millions of years), and that leaves time for organisms to evolve rather than go extinct. In contrast, the start of the PETM was geologically very fast. Within a few thousand years, the deep ocean warmed by about 5°C. It also became more acidic and less oxygenated. All these changes may have contributed to the demise of many species of forams, but most scientists who have studied these environments suggest that the abrupt temperature increase was likely the main cause of this extraordinary extinction.[15] One of the many species of deep-water forams that went extinct early in the PETM is *Cibicidoides velascoensis*, which is illustrated in figure 4.5.

Another important PETM change was the near-complete destruction of coralgal reefs (i.e., reef communities dominated by corals and coralline algae). This is best documented in the Tethys Sea (between Eurasia and Africa)—extending from the Atlantic coast of what is now Spain, to various locations in northern Africa, parts of what are now the Balkans, a location in central Asia, the Horn area of Africa, and northern and eastern India (figure 4.6)—but it likely happened elsewhere as well. Late Paleocene reef systems, which probably resembled the tropical coral reefs we are familiar with today, disappeared and were replaced by carbonate mudflats rich in forams. This change has been ascribed to a combination of high water temperatures, ocean water acidity, and an enhanced hydrological cycle, which resulted in increased introduction of sediments and nutrients into near-coast waters leading to local anoxia. All of these would have been detrimental to coralgal reef ecosystems.[16]

This was not an extinction event but an extirpation[17] of the reef organisms from where they had been living. Coralgal reefs continued to hang on in other places, and so they were able to re-establish in the Tethys Sea after the PETM.[18]

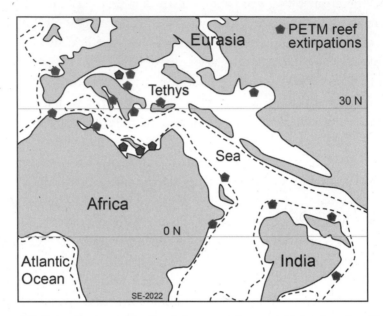

Figure 4.6: *Tethys Sea locations where Paleocene reef communities were replaced by foram mudflats.[19]*

It's worth thinking about this extirpation for a moment. Reef ecosystems, which are critical habitat for one-quarter of all marine organisms, were wiped out over a vast area for about 100,000 years. That may be just a geological instant, but in the context of human civilization, 100,000 years is effectively forever.

There is evidence of other types of extirpations along the Atlantic continental margin of North America. In Virginia and Maryland, marine rocks of Paleocene age have rich assemblages of molluscs (snails, clams, oysters, etc.), while almost no calcareous fossils are present in the overlying PETM beds.[20]

What of the larger organisms? They are more difficult to study because there are far fewer sharks, for example, than there are tiny benthic forams. We can assess patterns of corals and molluscs in well-exposed marine sedimentary outcrops on land in the Mediterranean region or the Gulf Coastal Plain, but with a few exceptions, the only way to study Paleocene deep-ocean sediments is by drilling into open-ocean seafloor sediments. Small fragments of large-animal fossils are only discovered by accident in ocean cores, so it's virtually impossible to assess differences in their abundance over short time periods. We do know that most types of marine organisms that were present in the Paleocene are still present today, so we can assume that there were few significant extinctions of large marine organisms during the PETM. In other words, while almost all marine species would have been severely stressed by the extreme conditions of the PETM—high water temperatures, anoxia, acidity, and a reduced food supply—most were able to find some place to make a living, and then were able to start the process of expansion and recovery about 150,000 years later.

Summary

The oceans were very warm during the PETM, on average 5° to 6°C warmer than during the late Paleocene (which was already warm). The warming of surface waters was greatest at temperate latitudes, and the least in the tropical Pacific Ocean. The warming extended to the deep ocean, which also warmed by around 6°C relative to the late Paleocene.

The pattern of deep-ocean circulation changed in two important ways during the PETM. In the late Paleocene, there was deep-water formation in the Southern Ocean. In the PETM, the deep-water formation moved to the tropical parts of the oceans, vertical circulation was less vigorous and did not extend as deep as it had previously.

Sea level rose during the PETM, although we don't know by how much. It might have been just a few meters, and that could be explained by the change in ocean temperature. But it might have been as much as 20 to 30 meters, in which case, melting of permafrost would have been necessary.

The oceans became more anoxic during the PETM, both at depth and in shallower shelf areas. The deeper anoxia was likely a result of reduced circulation. The shallower anoxia may have resulted from the higher water temperature and an enhanced hydrological cycle that brought more sediments, organic matter, and nutrients into coastal waters.

The PETM oceans were more acidic than those of the late Paleocene oceans, primarily because of higher atmospheric CO_2 levels. The greater acidity resulted in significant shallowing of the depth at which calcium carbonate becomes soluble (the CCD) and so changed the composition of seafloor sediments. The stronger acidity likely also had negative implications for organisms that made their hard parts out of calcium carbonate.

The ^{13}C proportion of carbonates in seafloor sediments dropped by between 0.1% and 0.3% during the PETM, most likely because of the release of a large amount of ^{13}C-poor carbon (e.g., from permafrost) to the atmosphere.

There was a massive and unprecedented extinction of deep-water forams during the PETM (30% to 60% of all species) likely due to warming of the deep ocean, although acidity and anoxia may have played a role. There was also near complete extirpation of coralgal reef communities from the Tethys Sea area. Again, the main factor may have been high water temperatures, but acidity and local anoxia are likely to have contributed. It's likely that marine life in these areas remained highly impoverished for at least 150,000 years.

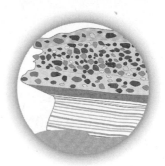

Chapter 5

Changes on Land During the PETM

Thus, global environmental events, such as the PETM, have had profound effects on evolution in the geological past and must be considered when modeling the history of life.

—Philip Gingerich[1]

A S WE'VE SEEN IN CHAPTER 4, there were major changes in the oceans during the PETM. If humans had been around, they might not have noticed the massive extinction of tiny organisms in the deep oceans, but many would likely have been dismayed by the dramatic destruction of reefs and other near-shore ecosystems because their food sources would have been decimated, and they would certainly have been affected by the extraordinary climate and ecosystem changes on the land. Not only would these have had an immediate impact on their day-to-day lives, they would likely have been repeatedly forced to make agonizing decisions—like tiny *Theilhardina* from chapter 2—to remain where they were and die or to pack up their stuff and move elsewhere.

The Terrestrial Rock Record of the PETM

About half of the seafloor is older than 55 million years. If you drill into the seafloor sediments in almost any location, you will eventually encounter the thin layer—in many cases under a metre thick—that represents the PETM. That's because sediments are accumulating almost everywhere in the oceans, albeit very slowly, almost all of the time, and also because seafloor sediments are quite stable. Once they accumulate, they get covered with more sediments, and then they remain, essentially undisturbed, for tens of millions of years.

The situation is very different on land. First, deposition is not continuous in most places; it is typically only happening in river valleys, deltas, and lakes. Second, most of those sediments get eroded and transported away before they can ever be turned into rock. Third, because of tectonic activity, many of the rocks on land are later folded, faulted, and uplifted to form mountains, and then massively eroded, resulting in the erasure of entire periods of geological information. So, as is shown in figure 5.1, the record of the PETM in terrestrial rocks (rocks deposited on land) is sparse and scattered. There are a few hot spots, including one that we've already visited in northwestern Wyoming, and another in western Europe, with exposures in Spain, France, Belgium, and southern England. There may be good exposures of terrestrial PETM rocks in many other places, but it will take lots of time and effort to find and study them. Some of the locations shown in figure 5.1 are actually sites with near-shore marine sedimentary deposits, either now on land or sampled through ocean drilling. Materials from these sites have been used to understand the conditions on land through studies of pollen and other terrestrial organic components.

Although PETM rocks are uncommon on land, they are much easier to access and to study than those on the seafloor, and in many cases, they can provide more and better information. For example, they may exist

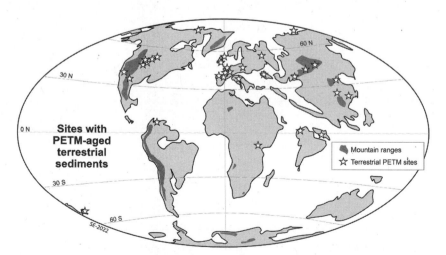

Figure 5.1: *Locations that have provided evidence of PETM terrestrial conditions.*[2]

in outcrops that extend across the terrain for hundreds or thousands of meters, and because sedimentation tends to be much more localized on land, they may be tens of meters thick—as opposed to less than one metre for most marine PETM layers. Because of this greater visibility and extent, there is a much better chance of finding macrofossils in terrestrial outcrops, including seeds, leaves and stems of plants, and teeth, bones, or even whole skeletons of animals. Because of their greater thickness, we can use them to delineate changing conditions in much finer detail than with the thin seafloor layers.

Climate Changes on Land

We know, from proxy estimations at several scattered locations, that the temperature on land increased significantly during the PETM, but using just that type of information to determine regional temperature-change patterns is questionable because of the limited number of terrestrial PETM sites. Another approach is to use a general circulation model (GCM) coupled with proxy estimates of local changes. The results of a recent study based on that technique are summarized in figure 5.2. The results indicate an average annual global air temperature increase of 5.6°C (for both land and sea areas), from 28.5°C in the late Paleocene to 34.1°C

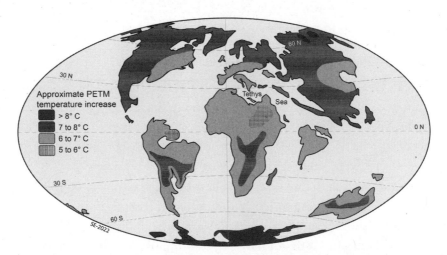

Figure 5.2: *Representation of terrestrial temperature increases between the late Paleocene and the PETM based on a GCM coupled with proxy temperature records.*

during the PETM. The model results also show that air temperature increases were significantly greater on land than over the oceans. The greatest increase was in Antarctica, at consistently more than 8°C. As the study authors explained,[3] this is likely due to the general tendency for polar regions to warm more significantly than other regions, and also to a modeled decrease in winter snow cover extent and duration in Antarctica. In most northern hemisphere land areas, the modeled temperature increase is between 7° and 8°C (although it exceeded 8°C in northern Asia), while in most parts of South America, Africa, India, and Australia, it is between 6° and 7°C. The approximate overall land-area temperature increase is close to 7°C.

The digital model output in Tierny et al. on which figure 5.2 is based also provides information about precipitation and evaporation changes. It indicates a strong increase in precipitation in equatorial regions and less precipitation in adjacent subtropical regions, along with a moderate increase in precipitation in most temperate regions, except in western Eurasia and the land surrounding the Tethys Sea, which would have seen a slight reduction. Moisture balance on land is significantly influenced by evaporation, and the higher temperatures of the PETM resulted in dramatically increased evaporation everywhere, which means that there was likely greater aridity over most continental regions, even in areas where there was more rainfall.

Paleocene and PETM Climate Types

We can visualize the late Paleocene and PETM climates by comparing them with the climate types on Earth today. Such a comparison, based on the Köppen Climate Classification,[4] is shown in figure 5.3. In the late Paleocene, the northern parts of the northern hemisphere and Antarctica were within what we describe today as "continental warm" with hot or warm summers and cool or even cold winters. Examples today would be parts of the midwestern US or the northern parts of France. In most such areas, snow would have accumulated over the winters, and in the more extreme parts, it might have stayed on the ground well into spring.

In areas a little closer to the equator, the climate would have been similar to what we currently call temperate wet or dry. Temperate wet

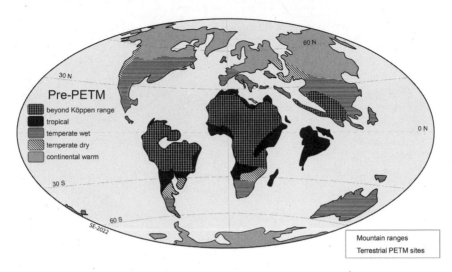

Figure 5.3: *Generalized distribution of Köppen climate types in the late Paleocene.[5]*
The patterns are derived from climate modelling and also based on information from
terrestrial pollen types found at many locations on land and at what would have been
near-shore ocean sites.

is humid subtropical, like the southeastern parts of US today, and in the
Paleocene, it would have applied across almost half of North America,
including parts of southern Canada, as well as much of southern Europe
and Asia. Temperate dry is like the climate of the Mediterranean region
or parts of California. In the late Paleocene, "tropical" climates extended
across parts of South America and Africa, and all of India. These areas
would either have been wet tropical, like much of equatorial Africa or
South America today, or dryer savannah climates, like those of subequa-
torial regions in Africa. Finally, most of central South America and
Africa and parts of southern Asia had climates that are hotter than any
found on Earth today, and these are shown as "beyond Köppen range."
That doesn't necessarily mean too hot for anything to grow, but likely
too hot for most plant types that we are familiar with. The hot late
Paleocene climate had evolved over tens of millions of years, so plants
and animals had plenty of time to evolve and adapt to the heat.

The modelled Köppen climates of the PETM are shown in figure 5.4.
With temperatures around 6° to 8°C warmer than the late Paleocene,

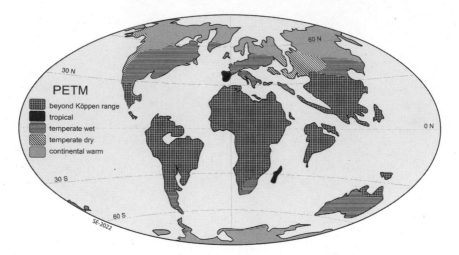

Figure 5.4: *Generalized distribution of Köppen climate types during the PETM.*

the climate zones have migrated poleward. The areas of warm continental climate in the far north and Antarctica have become smaller. The temperate wet and dry regions have also expanded poleward, reaching well into Canada, the northern parts of Europe, and parts of coastal Antarctica. These areas had PETM climates similar to what currently applies in the southeastern US. And, of course, the "beyond Köppen range" regions have also expanded, covering all of southern Asia and India, most of Africa and South America, and much of Central America.

The key point, of course, is that these changes did not take place over millions of years; they happened in just thousands of years. Plant communities could not fully adapt at that rate—although they likely did eventually—and animals were forced to move, as we have already seen, and will see more of below.

Examples of Specific Climate Changes

Specific features of PETM rocks in some areas provide evidence for climate changes that are not reflected in GCM studies like those described above. For example, in northern Spain, analysis of rock textures shows that what had been a gentle semiarid coastal plain in the late Paleocene was transformed, during the PETM, into a vast cobble and boulder field that extended over thousands of square kilometers.[6] The resulting

Figure 5.5:

The Claret
Conglomerate
(upper right) in
the foothills of
the Pyrenees,
northern Spain.
(Photo by Robert
Duller, University
of Liverpool. All
rights reserved.
Used with
permission.)

coarse-grained rock, known as the Claret Conglomerate (figure 5.5)—deposited by fast-flowing rivers over several thousand years—is assumed to be the result of an enhanced global hydrological cycle (more evaporation and more precipitation) that led to intense storm systems that most likely originated over the Atlantic Ocean. Several other locations have evidence of enhanced erosion and sedimentation that are consistent with a stronger hydrological cycle,[7] and this is consistent with the current trend of increasing storm intensity.[8]

In the Bighorn Basin of Wyoming, textural and chemical analysis of sedimentary layers, which are essentially preserved soil horizons, shows a strong trend towards drier conditions in the early part of the PETM. Those working on this problem[9] stress that the evident drying of the soils during the PETM may be a result of changes in the timing of precipitation as much as in the total amount of precipitation. Drying is also related to the well-documented increase in temperature. There were dramatic changes in the ecosystems of the Bighorn Basin, changes that correspond with the incidence of anomalously dry PETM soils.

Another key feature of the PETM rocks in Wyoming is their striking variation in color, from deep red and purple to dull grey (figure 5.6). The red and purple colours (dark layers in the black-and-white image) represent intervals of relative aridity—intervals lasting many thousands of

Figure 5.6: *The lower Eocene (PETM) Willwood Formation at the southern end of Polecat Bench in the Bighorn Basin, Wyoming.*

years. This pattern underscores the cyclic nature of climate fluctuations during the PETM.

Terrestrial Isotopic Records

For well over a century, paleontologists and geologists have been picking their way through outcrops in Wyoming's Bighorn Basin looking for fossils of trilobites, early reptiles, early mammals, dinosaurs, and many others, and of course for fossil fuels (coal, oil, and gas). That has included a lot of painstaking, hot, dusty, and sweaty hands-and-knees study of the rocks of the Paleocene and Eocene. In 2011, two core holes were drilled through the lower Eocene and upper Paleocene layers in the Polecat Bench area of the basin.[10] The holes were located about five meters apart, near the plateau area visible at the top left of figure 5.6. The roughly 300 meters of core from these holes has provided valuable information about the PETM, first because the core is continuous and the full record of the PETM doesn't have to be pieced together from many different sites—with a significant potential for missing important intervals—and second because the PETM rocks are deep enough at the location chosen that they are largely unaffected by the surface weathering that can alter the texture and chemistry of rock samples collected from outcrops.

The core from the Polecat Bench holes was sampled in detail (with about 180 samples over the 65 meters of the PETM interval) for analysis of isotopes and other chemical parameters. The results of the carbon-isotope analysis are shown in figure 5.7. The drop in ^{13}C levels, or carbon isotope excursion (CIE) is dramatic and deep—about a 0.7% difference. That compares with a decline of only about 0.2% in most seafloor sediment sections. The CIE is characterized by a precipitous decline at the start, a noisy period of low levels, and then a steady increase during recovery to near normal isotope levels. The CIE represents a change in the ^{13}C composition of the atmosphere that must have resulted from a massive release of carbon with very low ^{13}C levels, probably in the form of methane, either from permafrost and soil or from the deep ocean, or both.

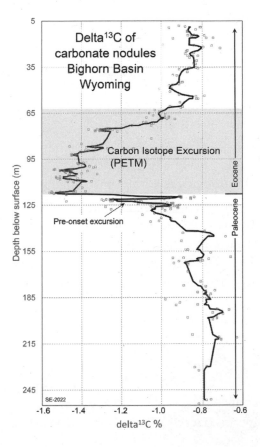

Figure 5.7: *The carbon isotope record for two PETM drill cores from the Polecat Bench area of Wyoming.[11]*

The extra information afforded by the detailed sampling from the Polecat Bench holes has allowed for definition of a separate short pre-onset isotope excursion (POE) prior to the main CIE. The POE developed over about 1,000 years and lasted for between 2,000 and 5,000 years. The period of more "normal" isotope values between the POE and the CIE also lasted a few thousand years, and the steep drop in ^{13}C values—the onset of the CIE—took about 1,000 years.[12] Both events are assumed to have been caused by release of significant amounts of methane into the atmosphere, and it is suggested that the second one (the main CIE) may have resulted from feedback mechanisms related to the first one (POE). The possible triggering and feedback mechanisms are discussed in chapter 6.

There is also evidence of a POE in data from at least one other location in Wyoming, from other terrestrial locations in China and Tibet,[13] and from two marine sections, one on the US Atlantic coastal plain (see figure 3.6) and one in the southern Atlantic near Africa.[14]

PETM Impacts on Land Plants and Animals

The rocks in the Bighorn Basin of Wyoming are the best studied PETM rocks anywhere—by far. This is true for several reasons. First, the PETM layers are exposed at surface over a large area of the basin; the outcrops are up to a few hundred meters wide along a linear extent of over 200 kilometers. Second, because of Wyoming's relatively dry climate, there is only light vegetation and sparse population in the region, making the rocks easy to see and to access (figure 5.6). Third, as already noted, this region of Wyoming has attracted paleontologists for well over a century—mostly because of the allure of oil and gas and of dinosaur remains in the Mesozoic rocks—so its geology and paleontology are well studied and well documented.

The importance of the early Eocene strata of the Bighorn Basin was understood long before the discovery of the PETM in 1991[15] because it was already known (since the early twentieth century) that the rocks contained fossils of the oldest primates in North America and also of early members of the ungulates, including the first artiodactyls (even-toed ungulates, like deer) and perissodactyls (odd-toed ungulates, like horses). It was also known that an early Eocene change in vegetation patterns was consistent with a significant increase in local temperature.[16] Here we look more closely at the ecosystem changes that took place in the Bighorn during the PETM, and then at evidence for changes at other terrestrial locations around the world.

There are lots of scientific names in this chapter, orders, families, genera, and species. Many of them will be unfamiliar to you and most are difficult to pronounce. I have tried to include more common names for organisms wherever possible, but in many cases, such as Condylarthra, Pantodonta, and Dinocerata, that's just not possible because these are obscure and extinct orders that don't have common names. Please be patient, and don't worry if the names won't stick in your head. If you come away with the feeling that many of these organisms are confusing, then you've actually got the message.

PETM Floral Community Changes in the Bighorn Basin

As noted above, the Bighorn area was within the temperate wet Köppen climate zone in the late Paleocene (figure 5.3). During the PETM, it was within the continental warm zone, and the local mean annual temperature was likely about 7°C warmer than it had been just a few thousand years earlier (figure 5.4). Although there might have been more precipitation overall, the enhanced evaporation due to the elevated temperature likely made it drier.

The floral assemblages of the Bighorn Basin before, during, and after the PETM have been described by Wing and Currano,[17] based on analysis of leaf, seed, flower, and pollen fossils from over 200 locations. The details of temporal changes in plant communities are summarized on figure A.2 in the appendix. During the late Paleocene, the forests in wetter sites of the Bighorn Basin were dominated by conifers, including redwoods and cypresses, along with cattails, ferns, and horsetails. Drier sites were populated with gingkos, katsuras, proteas, witch-hazel, and numerous other broadleaf trees (birch, maple, beech, elm, poplar, walnut, and dogwood).

Almost 90% of the late Paleocene flora of the Bighorn area were extirpated from the region during the PETM. This dramatic change was likely a result of the significantly increased temperature, relative drought, and substantially higher CO_2 levels. It is likely that stream flows generally declined and also became strongly seasonal and more variable (more flash flooding), and that permanent wetlands became rare or nonexistent. The PETM forests were more open than the Paleocene forests, if for no other reason than the loss of the evergreen conifers and the understory ferns and horsetails. It is important to point out that no known floral extinctions are directly associated with the PETM in Wyoming, only extirpations. Those extirpated species survived in other regions (likely towards the north or at higher elevations) and then repopulated the Bighorn after the PETM, when a climate similar to that of the late Paleocene was re-established.

The PETM immigrants to the Bighorn region included mulberry, laurel, magnolia, at least two trees in the pea family (similar to Mimosa), and sumac. Although other walnut genera existed both before and after the PETM, the genus *Platycarya* may have evolved at the start of the PETM.[18]

PETM Floral Community Changes in Other Regions

While there is no other location with a PETM terrestrial fossil record as detailed and well understood as that of the Bighorn Basin, there are rocks of PETM age in many other parts of the world (see figure 5.1), and some of these have evidence of significant changes to terrestrial ecosystems.

Pollen in near-shore marine sediments from New Zealand show a decline in conifers and ferns at the start of the PETM and an increase in broadleaf plants. There is evidence of a particular increase in the abundance of the rose family tree *Casuarina*.[19]

Pollen studies in the near-equatorial Columbia-Venezuela region show an increase in diversity during the PETM; most of the new genera were broadleaf plants.[20] As seen in figures 5.3 and 5.4, the climate of most of northern South America is classified as "beyond Köppen" both before and during the PETM. In spite of that, the authors of the Columbia-Venezuela study found no evidence that the elevated PETM temperature was beyond the tolerance of the tropical forest ecosystem in that region.

In northwestern Spain, a Paleocene forest dominated by conifers was replaced during the PETM by a relatively sparse and only seasonal community of broadleaf plants and ferns. Increased sediment transport from land is attributed in part to the reduction in vegetative cover.[21] The PETM climate was generally dry, with seasonal flooding and periodic catastrophic flooding.[22]

Pollen in near-shore marine sediments from the North Sea indicates a shift from a wetland conifer forest dominated by redwood and pine to a fern and broadleaf ecosystem dominated by alders.[23] The change was associated with enhanced input of terrestrial sediments to the marine system related to an increase in precipitation.

Finally, terrestrial pollen in marine sediments from the high Arctic north of Russia also shows a change from conifer-dominant to broadleaf-dominant ecosystems during the PETM, again with a warmer and wetter climate.[24]

All of these examples represent sites that were relatively close to a marine coastline during the PETM, and it would not be surprising, therefore, if none of them experienced the magnitude of climate change recorded at the Bighorn Basin, which was well over 1,000 kilometers inland from the Pacific Ocean. That doesn't mean that the profound floral changes in the Bighorn were unique in any way; most of the Earth's continental area was (and is) at least several hundred kilometers from the nearest coast, so it is possible that most land areas experienced floral changes on the scale of those in the Bighorn Basin.

PETM Faunal Changes in the Bighorn Basin

The Paleocene and Eocene vertebrate fossil record of the Bighorn Basin is remarkable for its abundance and completeness across a time span of several million years and also for its fine resolution at time scales of thousands of years. It is also remarkably well studied. As already noted, paleontologists have been working on these rocks for over a century, but the pace of study in the basin picked up significantly some 50 years ago, and much of that it is due to the insight and efforts of Philip Gingerich of the University of Michigan (see box 5.1 for more on Philip Gingerich).

Mammals radiated rapidly following the extinction of dinosaurs at the end of the Cretaceous (65 Ma), but early mammals are quite different from modern mammals, and it has proven difficult to understand how they fit into the categories that we are familiar with, such as carnivores, ungulates, rodents, and primates. The Bighorn Basin is highly significant for its abundance of Paleocene and Eocene mammal fossils. Several representative mammal genera of the Bighorn Basin are listed in figure A.3 in the appendix, which illustrates some of the changes in vertebrate fossil record during the late Paleocene and the early Eocene. Some late Paleocene mammals did not survive into the Eocene, and some of these disappeared from the region before the start of the PETM. But most Paleocene mammals did survive through the PETM and remained in the Bighorn area well into the Eocene. These include various primatomorphs, carnivoramorphs, condylarthras, pantodonts, and others (see table 5.1). For example, *Haplomylus* and *Ectocion* are ungulate-like condylarthras. *Tillodontia* has features that remind us of rodents, ungulates, and carnivores, but it isn't part of any of those groups. *Didymictus* and *Viverravus*

Table 5.1 Description of Bighorn Basin mammal groups listed in figures 5.8 and A.3

Group	Epoch	Description	Example species
Carnivoramorpha	Paleocene and Eocene	Predator but not Carnivora	*Viverravus gracilis*
Condylarthra		Ungulate-like	*Ectocion parvus*
Primatomorpha		Possible primate ancestor	*Plesiadapis dubius*
Tillodontia		Rodent-like	*Azgonyx grangeri*
Insectivore		Shrew-like	*Apheliscus nitidus*
Pantodontia		Herbivore	*Coryphodon eocaenus*
Rodent		Early rodent	*Parymus taurus*
Eulipotyphla		Hedgehog family	*Macrocranion junnei*
Perissodactyl	Eocene only	Odd-toed ungulate	*Sifrhippus sandrae*
Artiodactyl		Even-toed ungulate	*Diacodexis ilicis*
Primates		Arboreal primate	*Theilhardina brandti*
Multituberculata*		Rodent-like mammal	*Ectypodus tardus*
Hyaenadontidae		Predator but not Carnivora	*Prototomus deimos*

* Although only observed in the Eocene in the Bighorn Basin, Multituberculates are known from as far back as the Jurassic, some 100 million years earlier.

are predators, but cannot be classified with modern carnivores, as those did not evolve until much later in the Eocene. *Plesiadapis* and *Ignacius* are Primatomorphs (like primates, but not quite).

A total of fourteen different mammal genera (including the nine shown in figure A.3 of the appendix) are first observed in the PETM layers in the Bighorn Basin. This includes representatives of three orders—Perissodactyla, artiodactyla, and primates—that are not known to have existed anywhere on Earth prior to the Eocene. All three appeared almost simultaneously in Asia, Europe, and North America during the PETM. The Hyaenadontidae *Prototomus* is a predator, but cannot be classified with modern carnivores, which did not appear for about another 15 million years.

Gingerich[25] has suggested that these new mammals might have evolved over as little as a few hundred generations—a remarkably short time—in response to the pressure of dramatic climate change. As discussed in chapter 2, their nearly coincident appearance in Asia, Europe, and North America implies not only rapid evolution but relatively fast migration, likely also in response to the pressure of climate change.

Figure 5.8 provides a summary of mammal populations in the Bighorn Basin in the latest Paleocene and during the PETM in the earliest Eocene. The groups shown are described in table 5.1.

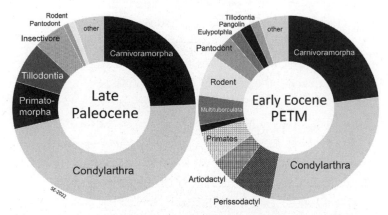

Figure 5.8: *Changes in mammal population proportions between the late Paleocene and the early Eocene (PETM) in the Bighorn Basin based on numbers of individuals in the fossil record. See table 5.1 for an explanation of the group names.*[26]

The main changes over a few tens of thousands of years at the start of the PETM are a significantly reduced proportion of Condylarthra, the emergence of Perissodactyla, Artiodactyla, and Primates, and the presence of a few other new taxa, such as Eulipotyphla and Pangolin. It is clear that, in spite of the significantly changed climate and vastly different vegetation mixture, the PETM was actually marked by increased mammal diversity in the Bighorn Basin. Relatively few Paleocene mammals disappeared, and many new Eocene mammals appeared. This may be due in part to the apparent rapid rate of evolution in response to the changed climate.[27]

Box 5.1 Philip Gingerich and the Bighorn Basin

Philip D. Gingerich is a Professor Emeritus in three departments at the University of Michigan: Biology, Geology, and Anthropology, and he was the long-time director of the University's Museum of Paleontology. Born in 1946, he studied at Princeton and Yale. He began teaching and research at Michigan in 1974.

Gingerich first studied the Paleocene and Eocene strata of the Bighorn Basin in the early 1970s; over 50 years later, he is still actively contributing to our understanding of its fossils and rocks and their implications for the PETM.[28] He has contributed to well over 100 publications on Bighorn paleontology, and he and his colleagues and students are responsible for the collection, description, analysis, and cataloging of thousands of fossils, and for defining the nature of Paleocene and Eocene mammal evolution in the basin, including the longer-term evolution over millions of years and the shorter-term evolutions triggered by the rapid changes of the PETM. The discovery of those evolutionary changes preceded the initial discovery of the PETM climate changes.

Notwithstanding his larger-than-life role in the Bighorn Basin, Philip Gingerich is actually best known for his work on the evolution of whales. There were never any whales in the Bighorn; those investigations have taken him to Pakistan, northern Africa, and the Middle East.

Non-mammalian fossils are relatively rare in the Bighorn Basin. Some of the more common ones include molluscs (freshwater bivalves and land snails), aquatic arthropods (crayfish), salamanders, reptiles, including turtles, lizards, crocodiles, and the crocodile-like *Champsosaurus*, plus a few birds.

Because of the scarcity of specimens, there is very little information on changes in composition of the non-mammals across the Paleocene-Eocene boundary. The exceptions are *Champsosaurus*, which arose during the Cretaceous (and survived the dinosaur extinction event) but then disappeared at the end of the Paleocene, and the large ground-dwelling bird *Gastornis*, which appears only to have been present in the Bighorn Basin during the Eocene.[29]

In addition to the observed changes in the composition and diversity of fauna in the Bighorn, there is evidence of some striking changes in the sizes of individual species. Figure 5.9 shows the variations in the surface area of the first molar in *Sifrhippus* (the first known horse) from the time of its appearance at the start of the PETM through to at least 100,000 years after the PETM. Tooth size is proportional to body size in most organisms, and this change is estimated to represent a 39% reduction in body mass of *Sifrhippus* during the PETM.[30] When it first evolved, *Sifrhippus* was about the size of a fox (figure 1.8). By the end of the PETM, it was no bigger than a house cat.

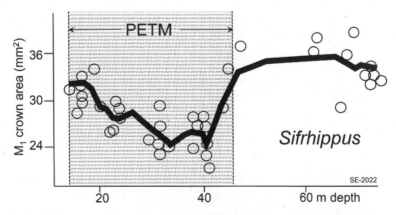

Figure 5.9: *Variations in the surface area of the first lower molar (M1) in* Sifrhippus *during and after the PETM.*[31]

There is evidence that other animals experienced similar dwarfing, including the condylarths *Ectocion parvus,* 44% smaller, and *Copecion davisi,* 40% smaller. And it's not just mammals that shrank. The soil burrows left behind by insects and worms got smaller by between 30% and 50% during PETM.[32]

Bergmann's rule[33] holds that animals tend to be generally smaller in warm climates and larger in cool climates, and that this is partly related to the body's surface area to mass ratio. Animals need to shed body heat during hot conditions, and that is easier for small animals than it is for large ones. Gingerich argues that body size might also be related to food quality and quantity,[34] and there are good reasons to expect that both of these decreased during the PETM. The elevated atmospheric CO_2 level of the PETM likely played a role in this regard since higher CO_2 levels lead to reductions in nutrient levels in plant matter, and as we've seen, the relatively arid conditions in the Bighorn during the PETM resulted in a lower density of vegetation than was the case before and after.

The late Paleocene and early Eocene terrestrial faunal fossil records at other locations around the world are not sufficiently detailed, or have not been studied in sufficient detail, to make inferences about changes related to the PETM.

Summary

The terrestrial rock record of the PETM is quite different from the marine rock record. While marine PETM layers are distributed evenly across most of the seafloor, terrestrial PETM layers are relatively rare. Where they are found, they tend to be significantly thicker than their marine counterparts. They are also much easier to sample and study, especially in places where they are exposed in outcrops.

Evidence from terrestrial rocks indicates that there was significant climate warming on land during the PETM. The average land air temperature rose by about 7°C with more pronounced differences in polar regions than in equatorial regions. Changes in precipitation were unevenly distributed, but while many land areas received more rainfall overall, the higher temperatures resulted in stronger evaporation, so there was less moisture availability in most areas.

As would be expected, modelling shows that the various climate-type regions were situated closer to the poles during the Paleocene than they are now. They moved even further poleward during the PETM, at which time almost half of the Earth's land areas had climates that were hotter than any known climate on Earth today.

A series of extensive conglomerate layers in northern Spain provide evidence that the region was subject to anomalously strong storm activity during the PETM.

Samples collected from two holes drilled through the PETM layers in the Bighorn Basin of Wyoming provide a uniquely detailed record of the carbon isotope excursion (CIE) and show evidence of a short pre-CIE onset excursion (POE). The POE developed over approximately 1,000 years. The CIE started several thousand years later, and that transition also took about 1,000 years. It is possible that the CIE resulted from feedbacks initiated by the POE.

The PETM climate changes resulted in striking floral community changes. In the Bighorn Basin, there is evidence of extirpation of 88% of plant genera (including all conifers) and immigration of several new genera—likely from southern latitudes—especially members of the pea family. Extirpations of conifers was also recorded in many other locations around the world.

Faunal extirpations in the Bighorn Basin were not as dramatic. There were several immigrant genera and some important new genera, including Perissodactyla, Artiodactyla, and Primates, leading to an overall increase in faunal diversity. There is consistent evidence for dwarfing of some taxa, including a perissodactyl, two condylarths, and some soil-dwelling organisms.

In case you're not convinced about the significance of the near-simultaneous first appearance of Perissodactyla, Artiodactyla, and Primates at the very start of the PETM, have a glance at figure 5.10. That's a primate, of course, along with two other animals that have been critically important to humans during historical times.

Figure 5.10: *An artiodactyl, a perissodactyl, and a primate. (Photo by Jean Beaufort, publicdomainpictures.net.)*

Chapter 6

What Caused the PETM Runaway Climate?

There is still a great deal of mystery about the PETM—the trigger, where the carbon came from and what happened to it—and the latest research hasn't tied up all the many loose ends.

—Gavin Schmidt[1]

T HE PREMISE OF THIS BOOK, which is supported by the work of dozens of climate scientists, is that the PETM runaway climate event is a credible analogue for our future climate. In other words, it is possible that something similar could happen in our time, in this case triggered by the massive and rapid anthropogenic release of carbon into the atmosphere. The key question we need to ask, therefore, is what caused the climate to change so dramatically 56 million years ago? To answer that question, we have to consider the potential sources of carbon, the event and mechanism that likely first triggered the release of carbon to the atmosphere, the feedback processes that would have accelerated the warming by releasing even more carbon, and then, finally, the mechanisms that gradually pulled the Earth back into climate "normality" after about 180 thousand years.

Before we delve into those topics, we should first go back through some of the information that there was a PETM runaway climate in the first place, how much and how quickly conditions changed, and how life was affected. We have a number of crime scenes, with very old evidence—some of it ambiguous—and we have some suspects, but we have no living witnesses, and certainly no motive.

A Brief Review of the Evidence

The evidence presented above is that:

- the Earth's surface air temperature increased by about 7°C over as little as 1,000 years at the start of the PETM, implying a sudden increase in greenhouse gas (GHG) levels;
- ocean water also warmed by nearly as much, even deep-ocean water;
- the carbonate compensation depth (CCD) of the oceans shallowed because the oceans had become more acidic due to the higher CO_2 level in the atmosphere;
- the $^{13}C/^{12}C$ proportion of carbon in the atmosphere and oceans dropped significantly because of the release of carbon rich in ^{12}C;
- sea level rose marginally; and
- life on the continents and in the oceans was seriously impacted—more in some places than in others.

Based on oxygen isotopes and other proxies, we know that the change in air temperature ranged from about 5° to 8°C, with more warming on land than over oceans, and more warming towards the poles than in tropical regions. Many regions received more rainfall overall, but the higher temperatures resulted in stronger evaporation, so there was likely less moisture available in most areas. There is also evidence of increased intense storm activity in some areas. The degree and speed of the warming can only have been a result of an increase in GHG levels, roughly equivalent to a doubling of the pre-PETM level.

Ocean sediment cores show us that the CCD (the water depth below which calcium carbonate dissolves) shallowed by as much as 2,500 meters in the Atlantic Ocean during the PETM (see figure 4.5) and by about 500 meters in the Pacific. The shallowing observed is consistent with a decrease of about 0.35 pH units, which was caused by an initial carbon pulse to the atmosphere of between 2,000 and 3,000 billion tonnes over a 1,000-to-5,000-year period, followed by a slower release of a similar or greater amount of carbon over the next several tens of thousands of years.[2] Modelling studies of how the Earth's average temperature, ocean acidity, and other factors will change as a result of the addition of thousands of billions of tonnes of carbon to the atmosphere consistently show that a short (ca 1,000-year) intense input of carbon would have had a much more dramatic effect than a longer less intense input of the same amount of carbon.[3]

The ^{13}C proportions of foram tests in ocean sediment cores dropped sharply in the early part of the PETM, remained low for most of the event, and recovered slowly towards the end. The total carbon isotope excursion (CIE) was between 0.1% and 0.7% ^{13}C. It was least in marine sediments—typically in the 0.1% to 0.3% range—and highest in terrestrial sediments. For example, there was a 0.7% drop in the Bighorn Basin, and between 0.5% and 0.6% at locations in China and Tibet. At the Bighorn Basin Polecat Bench drill site, there is evidence of a brief (2,000-to-5,000-year) isotope excursion (the pre-onset excursion) a few thousand years before the start of the main CIE (figure 5.7).

Several studies show evidence of a PETM rise in sea level of up to 5 meters, while two studies imply a rise of 20 or 30 meters. An increase of around 5 meters could have been caused by the observed thermal expansion of ocean water. A more significant rise would also have required melting of a significant amount of permafrost.

Some 30% to 60% of deep-water foram species became extinct during the PETM, most likely due to warming of the deep ocean. There was also very extensive extirpation of coralgal reef communities, likely also attributable to high water temperatures. Molluscs were also extirpated from many continental shelf areas. There were pronounced changes in vegetation types in may regions in response to higher temperatures and drier conditions. Coniferous forests were replaced by deciduous forests, and many plant types were extirpated, and replaced by immigrants from closer to the equator. A few land animals went extinct, but several new types evolved, including primates, artiodactyls, and perissodactyls. Several genera became significantly smaller for the duration of the PETM.

Paleocene Carbon Reservoirs

We need to start by considering the potential sources of carbon in the late Paleocene because without the addition of some carbon dioxide and/or methane to the atmosphere, it's impossible for the climate to change in the way that it did. Earth's major *present-day* carbon reservoirs are illustrated in figure 6.1. The area of each circle is proportional to the mass of carbon stored in that reservoir, and the numbers are also shown. It is likely that these proportions were generally similar during the Paleocene, although the methane hydrate (see box 6.1) reservoir may have been smaller because of higher ocean temperature. On the other hand, the permafrost reservoir may have been larger because of the large areas of Antarctica and Greenland that were not glaciated and there had been ample time for permafrost to accumulate.[4] DeConto and coauthors believe that the late Paleocene permafrost carbon reservoir could have been as high as 3,700 billion tonnes.[5] Some of the enormous amount of carbon in the crust is present at very low levels in crystalline rock (e.g., granite), while the rest is present in higher concentrations in sedimentary rock (e.g., limestone, organic-rich shale), and in unconsolidated sediments on the seafloor. That crustal reservoir also includes magma, and volcanic eruptions do release carbon dioxide. The circle that represents coal, oil, and gas is also actually part of the crust, and the same could be said for methane hydrate (see box 6.1).

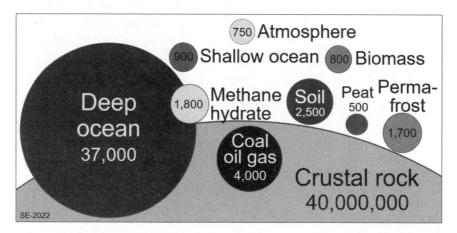

Figure 6.1: *Representation of the present-day relative sizes of the Earth's important carbon reservoirs; numbers are in billions of tonnes of carbon. (Based on data from several sources.)*

Box 6.1 What Is Methane Hydrate?

Methane is a gas with the formula CH_4. It is the main component of natural gas and is a potent greenhouse gas. Methane hydrate (aka methane clathrate) is an ice-like solid that consists of methane molecules trapped inside cages of water molecules (see the figure). It forms at low temperatures (close to 0°C and colder) and at relatively high pressures, similar to the pressures on the ocean floor. If allowed to break down at surface, a cubic meter of methane hydrate would release about 170 cubic meters of methane gas.

Ocean floor sediments include organic matter. As the sediments accumulate to tens or hundreds of meters thick, the older deeper sediments get warmed by heat flowing out of the Earth's interior. Bacteria and other microorganisms start digesting the organic matter, and they emit methane gas that slowly percolates up through the sediment pile. Near to the interface with seawater, the temperature is low enough for hydrate to form, and so most of the methane gets trapped there as methane hydrate.

There is a massive amount of methane hydrate within sediments on the seafloor around the world—approaching the energy equivalent of all known conventional coal, oil, and gas reserves. It is safe there as long as the deep-ocean water remains cold and greedy fossil fuel companies don't start trying to extract it.

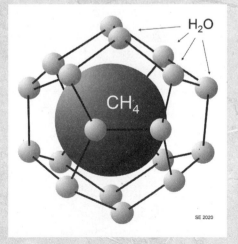

Under *stable* climate conditions, and without the intervention of humans, most of these reservoirs are effectively irrelevant because the carbon in them is not going anywhere—it is sequestered. That applies to the solid rock of the crust, of course, along with its contained magma, organic-rich shales, and fossil fuels. It also applies to the deep ocean, soil, peat, permafrost, and methane hydrate. Only biomass, the atmosphere, and the shallow ocean (and fresh water) play a significant role in day-to-day carbon cycling. Volcanism can make a difference, especially when it happens on a large scale.

But, under *changing* climate conditions, the carbon in permafrost, peat, and soil may become unstable and may be released into the atmosphere, especially if it gets warm enough to melt permafrost or if vegetation patterns change. In circumstances of exceptional climate change, the oceans could warm enough to destabilize the ocean-floor methane hydrate reservoir, especially along the continental shelves.

Of course, we humans have changed everything by rapidly releasing massive amounts of carbon from fossil fuels.

It seems that the climate system was ticking along in a stable way during the late Paleocene, with regular and normal exchange of carbon amongst the biosphere, the atmosphere, and the shallow ocean, but then something remarkable happened. Quite quickly (within as little as 1,000 years), the CO_2 level of the atmosphere increased dramatically and the climate warmed, and then it warmed a lot more. The potential and plausible sources of that CO_2 are melting and destabilization of permafrost, breakdown of soil and peat, destruction of biomass, destabilization and release of seafloor methane hydrate, and oxidation by volcanism of organic carbon stored in seafloor sediments. We'll look at each of these in more detail below.

Changing the climate normally involves both a trigger (an event that gets it started) and then feedbacks (processes that maintain and accelerate it). At present, the trigger is the massive anthropogenic release of fossil fuel carbon, and the feedbacks include melting sea ice and glaciers, destabilization of permafrost, and changing forests. There must have been a trigger, or perhaps multiple triggers, at the end of the Paleocene.

Potential PETM Climate Change Triggers
Volcanism

Volcanic activity can act as a trigger for climate change, either directly or indirectly. The direct changes are related to the gases released during a volcanic eruption, which, as we've seen in chapter 2, can lead either to short-term cooling, because of sulphate aerosols that block incoming sunlight, or to slow warming, because of the buildup of CO_2, if the volcanism continues at an elevated rate for at least tens of thousands of years. If the cooling is strong enough, it might trigger cooling feedbacks (snow accumulation, for example) that lead to more cooling. The same applies to warming, but as noted above, volcanism-related warming tends to be slow. The PETM warming was geologically fast—too fast.

Volcanism could also affect the climate indirectly in situations where it takes place on the seafloor and if magma passes through carbon-rich seafloor sediments, leading to the rapid oxidation and release of some of that stored carbon as CO_2.[6] There was ongoing volcanism in the North Atlantic during the Paleocene and Eocene (from approximately 63 to 52 Ma[7]), related to the opening of this part of the Atlantic basin (figure 6.2), and there is evidence from volcanic ash that reached a site on Denmark of two potentially significant eruptions just before the start of the PETM. These events could have been triggers for the PETM warming because the magma is assumed to have passed through organic-rich seafloor sediments and, in doing so, could have heated any contained organic matter converting it into CO_2.[8] Any methane hydrates stored in those sediments would also have been destabilized under these conditions. A recent study by Berndt and coauthors[9] has provided some support for the volcanic trigger concept, showing that, in some North Atlantic areas, volcanism took place in water shallow enough (likely less than 100 m) that any methane released during the process would have vented directly to the atmosphere, rather than oxidizing to CO_2 on its way up through the ocean water. This is significant because methane is a much more potent GHG than CO_2.

The layers of volcanic ash below the PETM clay, shown on the right-hand side of figure 6.2, are interpreted to represent windblown ejecta from two significant eruption events in the North Atlantic. Based

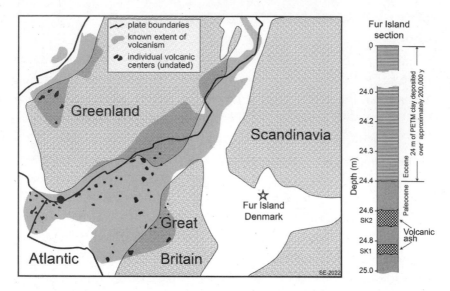

Figure 6.2: *Sites of Paleocene and Eocene volcanism in the North Atlantic and position of volcanic ash layers immediately beneath the PETM section at Fur Island Denmark.*[10]

on the sedimentation rate for the 24 meters of PETM clays at that location (~0.13 mm/y), and assuming a roughly similar rate for the 20 centimeters of clay immediately below the PETM, it is possible that the upper volcanic ash event took place between 1,000 and 2,000 years before the start of the PETM.[11] There is no evidence in the Fur Island rock layers of significant North Atlantic volcanism *during* the PETM.

Orbital Cycles

Milankovitch orbital cycles (see figure 2.6 and related text) can trigger climate changes—they certainly have during the past million years of northern hemisphere glaciation, and they are also known to have controlled monsoon cycles at earlier times—so it's reasonable to think that they could have been a climate factor during the late Paleocene.

In a 2019 study, Zeebe and Lourens[12] analyzed color reflectance data from Ocean Drilling Program (ODP) core from site 1262 (in the Walvis Ridge area of the South Atlantic) and compared the color variations—which they describe as being mostly due to changes in the proportion of white calcium carbonate—to the patterns of orbital

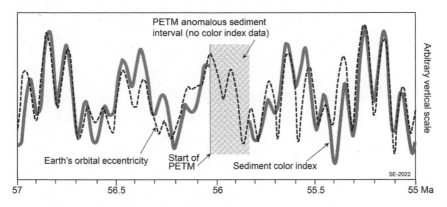

Figure 6.3: *Seafloor sediment color index values (heavy line) and orbital eccentricity values (dashed line) for the period from 57 to 55 Ma.*

eccentricity. This is illustrated in figure 6.3. There is a remarkably good match between the eccentricity cycles and the color-index patterns (except for the PETM interval during which the sediments have virtually no carbonates at all). Based on this correlation, the authors were able to show that the start of the PETM coincides almost exactly with a peak in the eccentricity curve. This supports the concept that a change in the climate related to orbital forcing could have triggered the PETM, although its difficult to argue that this was the only mechanism, since there were more extreme eccentricity peaks in the pre- and post-PETM parts of the record. This analysis also shows that the extent of the PETM—based on the interval with anomalous low carbonate levels—is about 180,000 years.

Zeebe and Lourens only note the coincidence between the eccentricity maximum and the start of the PETM and do not speculate as to how that coincidence could have resulted in a change that might have triggered the PETM.

Extraterrestrial Source

It has been proposed that there is evidence for an extraterrestrial (cometary) trigger for the PETM,[13] including an increase in iridium levels[14] and the presence of magnetic nanoparticles in pre-PETM sediments. The proponents suggest that a 9-kilometer diameter comet would have supplied enough additional carbon (~150 billion tonnes) to warm the climate quickly (and also create a carbon isotope excursion) and would have started extensive wildfires, releasing more carbon. Others argue against this idea, noting that the iridium anomalies are small and region-ally inconsistent and are most likely to be related to volcanic eruptions, that the magnetic nanoparticles could be of biological origin, and that a 9-kilometer diameter comet would almost certainly have left some unequivocal evidence of its collision with the Earth.[15]

Discounting the unlikely extraterrestrial source, the most credible options for triggering the PETM are the volcanism in the North Atlantic basin and Milankovitch cycles, both of which appear to have a tenable connection to the start of the PETM. It is quite possible that both are implicated. For example, the orbital configuration might have set the stage for strong warming, while the volcanic event opened the curtain. Similarly, although there were numerous volcanic eruptions in the several tens of thousands of years prior to the PETM, the eccentricity may not have been high enough for any of them to trigger a climate change. It is also possible that neither of these explanations is correct.

Sources of Carbon

The various reservoirs of carbon that would likely have been available as climate change feedback sources in the late Paleocene are shown in figure 6.4. The reservoirs are arranged in order from most available to be released as feedbacks in response to climate warming, at the top, to least available at the bottom.

At present, the Earth's permafrost carbon reservoir is approximately 1,700 billion tonnes, but as noted above, it may have been significantly larger—as much as 3,700 billion tonnes—in the late Paleocene. The methane hydrate reservoir, on the other hand, may have been smaller during the Paleocene than it is now because of higher ocean temperatures then. There are no strong reasons to believe that the other reservoirs (peat, biomass, soil) were significantly different in size in the late Paleocene compared with now.

The numbers in brackets on Figure 6.4 are the approximate ^{13}C levels of the carbon in each of these reservoirs. The lightest carbon, with delta

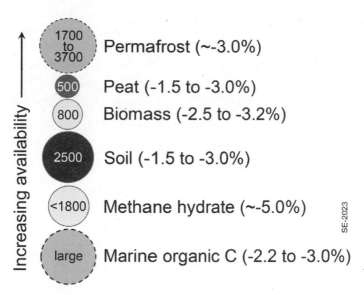

Figure 6.4: *Carbon reservoirs that likely existed at the start of the PETM and that could have been released to drive climate feedback processes. The numbers inside the circles are billions of tonnes of carbon. The numbers in brackets are the ^{13}C isotope levels in these reservoirs. (Based on data from several sources. See figure 3.9 and related text for more information on ^{13}C proportions.)*

^{13}C of around -5%, is in the methane hydrate reservoir. Most of the other reservoirs on the list have delta ^{13}C in the -2% to -3% range. The ^{13}C levels matter because the mechanism responsible for the PETM has to account not only for the observed warming and ocean acidification but also for the pronounced carbon isotope excursion observed in both marine and terrestrial sediments at sites around the world.

PETM Climate Feedbacks

A climate feedback is a process through which warming leads to more warming, or cooling leads to more cooling. At present, a strong feedback is melting sea ice in the Arctic Ocean. As the ice melts, the surface changes from being highly reflective snow-covered ice to nonreflective open water. Much more solar energy is absorbed by the water than could be absorbed by the snow and ice, and the area warms more, leading to the melting of more ice, and so on.

During the PETM, it is likely that methane stored in permafrost was the most readily available feedback reservoir (figure 6.4). With the initial warming that was triggered by an eccentricity maximum or by volcanism, or both, or by something else altogether, and with the greatest temperature changes at high latitudes, it is very likely that some of the permafrost would have started to melt, releasing its methane. This would have led to more warming, and destabilization of more permafrost, and that climate feedback could have continued for centuries until there was no permafrost left to melt.

Peat is second on the list of reservoirs because, presently at least, most deposits form within or near permafrost regions, and in some cases, their stability is dependent on the continued existence of the permafrost. Peat deposits are also vulnerable to destabilization and then oxidation on drying, and we know that some areas were drier during the PETM than they had been previously.

Biomass is third on the list. As we've seen in chapter 7, significant vegetation changes occurred during the PETM and there was likely less biomass overall than in the late Paleocene. Reduction in vegetation density could also have led to destabilization of soil, making the carbon in that reservoir more vulnerable to release. Widespread wildfires are also a possibility under the hot and dry PETM conditions.

Methane hydrates are low on the list because most are present in deep water, and also buried to depths up to tens of meters within the seafloor sediments. That means that they could not have been released until the deeper parts of the ocean warmed, likely thousands of years, and not until that warmth penetrated to sufficient depth in the sediments, additional thousands of years. Of course, it could have eventually happened, because the deep-ocean water did warm significantly, but it would have been progressive, with shallower continental shelf hydrate deposits going first and deeper deposits going last.

Finally, marine organic carbon is at the bottom of the list because it can only be released by seafloor volcanism, and while that did happen in the Paleocene and Eocene, including a known event just before the PETM that might have been the trigger (figure 6.2), there isn't strong evidence for significant volcanic events specifically during the PETM.

All the carbon reservoirs considered here have low ^{13}C values, and so all could have produced the ^{13}C anomalies (the carbon isotope excursion or CIE) observed in both marine and terrestrial records. It is appealing to think that we could determine which reservoir was the main culprit based on the differences in their ^{13}C values, but that is difficult because some reservoirs have similar ^{13}C levels and because we don't know the timing, the rates, or the magnitude of carbon releases during the PETM. All these variables can affect the end result. We can get some help from the geological records, including two that we have already looked at, as reproduced in figure 6.5. The calcium carbonate levels from three cores on the Walvis Ridge show us that the depth at which calcium becomes soluble (CCD) rose quickly in the first few thousand years of the PETM—to nearly 1,500 meters water depth—indicating a large pulse of carbon that produced fast and dramatic increase in ocean water acidity, and that the CCD then dropped gradually, remaining above 2,600 meters for several tens of thousands of years and above

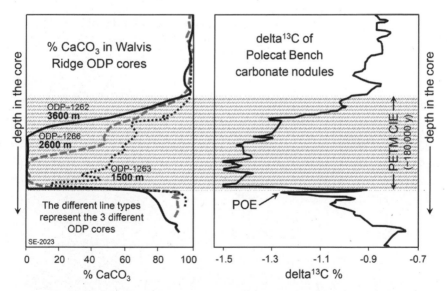

Figure 6.5: *Calcium carbonate concentrations in ODP cores from Walvis Ridge (left) and delta ^{13}C levels of carbonates from the Polecat Bench holes in the Bighorn Basin (right) (POE: pre-onset excursion; CIE: carbon isotope excursion). The bold numbers are the water depths at the three different sites.[16]*

3,600 meters for over a total of 100,000 years, before slowly dropping further.

The Polecat Bench carbon isotope record shows a pre-PETM isotope excursion a few thousand years prior to the main excursion (the POE in figure 6.5). The main excursion starts with a steep drop by about 0.6% over around 1,000 years, followed by a noisy (i.e., episodic) record over the next roughly 90,000 years, indicating that there might have been several carbon release events.[17] That is followed by a gradual, but also episodic, recovery over the last 90,000 years. The isotope record from a terrestrial deposit in China is similarly episodic.[18]

The implication, therefore, is that a series of releases of carbon started a few thousand years before the main PETM onset and continued through about 150,000 years of the PETM. It appears that the main onset pulse was the strongest. We don't know the source of the PETM carbon, and it is possible that there were multiple sources. Those might include permafrost, peat, biomass and soil, and eventually, methane hydrates, and the delays in activation of these different reservoirs could simply have been due to the time required for changes (e.g., temperature increases) to propagate through the various Earth systems.

The amount of carbon released to the atmosphere to produce the PETM records has been estimated at between 2,000 and 10,000 billion tonnes, with a median value of around 4,000 billion tonnes.[19] The range is wide because there are several potential sources and many options regarding timing and the degree to which the release was episodic. The top four reservoirs shown in figure 6.4 total 5,500 billion tonnes of carbon. If we include the methane hydrate reservoir, it could be as high as 7,000 billion tonnes.

What Brought the PETM to an End?

Positive climate feedbacks are described above as being critical to the rapid onset and long extent of the PETM, but it is very likely that negative feedbacks are what eventually brought it to an end. Some potential candidates are increased surface ocean biological productivity consuming CO_2, enhanced rock weathering also consuming CO_2, and increased albedo in continental areas because of thinning of forests.

Surface ocean biological productivity results in CO_2 being drawn from the atmosphere by planktonic organisms and converted to calcium carbonate shells and carbohydrate solids that sink to the seafloor and are sequestered there. Under warm PETM conditions and a high atmosphere CO_2 level, surface ocean microorganism productivity might have increased along with the rate of carbon sequestration, gradually drawing down the CO_2 level and eventually leading to cooling. One study of seafloor sediment geochemistry supports this concept and implies that the process was working throughout the PETM,[20] while a more recent study has shown that the effect was delayed until about 70,000 years after the start of the PETM[21] but could have been important as a negative feedback from then on.

Weathering of silicate minerals within rocks (especially the ubiquitous feldspar) results in conversion of atmospheric CO_2 to dissolved HCO_3^- ions, which are then transported to the ocean and converted either chemically or biologically into calcium carbonate that is sequestered on the seafloor. (See the appendix, section 1.) Since weathering rates increase with temperature and also with CO_2 concentrations, and so would have been higher than normal throughout the PETM, weathering was likely a very important negative feedback process. This might also have been enhanced by the strong hydrological cycle of the PETM.

Albedo determines how much of the sun's light is converted into heat on the Earth's surface or is reflected into space without warming the Earth. Areas of sparse forest and savannah are much more reflective (have a higher albedo, see chapter 2) than areas of dense forest with nearly closed canopies. If the Bighorn Basin is representative of mid-continent regions, it is likely that many land areas became more reflective during the PETM. That negative feedback would have contributed to cooling. The flip side of this is that the sparser forests would have consumed less CO_2; that might have countered the albedo effect.

Summary

Sedimentary layers deposited around the time of the PETM provide evidence for pronounced warming of the atmosphere and oceans, acidification of the oceans, shallowing of the carbonate compensation depth

(CCD), a distinct carbon isotope excursion (CIE), and fossil records of extinctions, extirpations, and important evolutions.

A massive transfer of carbon to the atmosphere is needed to cause a climate change like the PETM, and in the late Paleocene, carbon reservoirs on Earth were quite similar to those of today, including rocks with organic matter, fossil fuels, seafloor methane hydrates, permafrost, peat, soil, biomass, and the oceans.

Some sort of trigger is needed to start a significant climate change event. The leading candidates in the late Paleocene were volcanism in the North Atlantic leading to release carbon stored on the seafloor and changes in Earth's orbital configuration (Milankovitch cycles). There was a volcanic event in the North Atlantic a few thousand years before the start of the PETM, and there was an eccentricity maximum that coincided almost exactly with the start of the PETM.

Once triggered, a climate change event can be enhanced and then perpetuated by positive feedback mechanisms that supply additional carbon to the atmosphere. Source candidates for that carbon include melting permafrost, destabilization of carbon stored in peat and soil, reduction in terrestrial biomass, and destabilization of seafloor methane hydrate due to deep-ocean warming. It is apparent that releases of carbon were episodic. That might be because there were multiple carbon sources and because it takes varying amounts of time for feedback mechanisms to start working.

It is likely that the PETM was eventually and gradually brought to a conclusion by negative feedbacks, such as weathering of silicate rocks on land and enhanced productivity in the shallow oceans, both of which would have led to sequestration of carbon on the seafloor. Another possible mechanism is vegetation-related changes in albedo of land surfaces.

Chapter 7

What Is Similar Now and What Is Different?

The difference, the extra radiation that's absorbed [due to melting sea ice] is, from our calculations, the equivalent of about 20 years of additional CO2 being added by man.

—Peter Wadhams[1]

T HE MAIN POINT OF THIS BOOK is that it is critical for us to understand whether ongoing anthropogenic climate change might just be really miserable for everyone and deadly for millions of people, or if it has the potential to lead us into something much worse, a catastrophic runaway climate crisis like the PETM. To do that, we need to assess how today's Earth is similar to that of the late Paleocene, and how it is different, and then consider the implications of those similarities and differences in determining our possible future climate trajectory.

What Is Similar Now?

Although the Earth's plates are always moving, most at least a thousand kilometers since the Paleocene, the continental arrangement is not drastically different now from what it was then. As now, the Atlantic and Pacific Oceans were separated by North and South America on the one side and Eurasia and Africa on the other, although the Atlantic is now wider and the Pacific narrower. The significant concentration of land in the northern hemisphere was generally similar to what exists now, as was the total area of continental crust. Although the differences in continental positions are not that great, they have changed ocean currents, as we'll see below.

As described in chapter 6, there were massive carbon reservoirs in permafrost, peat, wetlands, and vegetation in the late Paleocene, and methane hydrate was stored within seafloor sediments. There was likely more permafrost carbon then, possibly more than twice as much, but less methane hydrate carbon.

The Paleocene ocean was a sink for carbon from the atmosphere and the biosphere, just as it is now. As atmospheric carbon was stored in plants on land and in marine microorganisms, it was gradually transferred to the deeper ocean.

What Is Different Now?

Ice, lots of it, and so many humans!

Ice

The current existence of glacial ice on the continents and floating ice on the polar oceans has profound implications for our climate and our climate future and, for many of us, the land that we live on and the water that we drink. Although we are in an interglacial hiatus, we are still in a glacial age, and although most glacial ice and sea ice is far away from where most of us live, it still has significant implications for our current climate and huge implications for our climate future.

At present, glacial ice covers roughly 18 million square kilometers, or 12% of the Earth's land surface. Should it all melt, its total volume is enough to produce at least 65 meters of sea-level rise. It is important to recognize, however, that some of the glacial ice is as much as 4 kilometers thick; it won't melt overnight. Sea-level rise is already an issue for millions of people, but the full 65 meters of sea-level rise could not happen for several thousand years.

Sea ice is the ice floating on the ocean surface in polar regions. Its area varies from season to season, but it currently covers about 25 million square kilometers on average, or 7% of the Earth's ocean surface. The melting of floating sea ice does not have sea-level implications.

Several future climate implications of glacial and sea ice will come into play as the Earth continues to warm. Of course, the degree to which we allow it to warm over the next several decades—and the rate of that warming—will affect the outcomes. The main one is the change in albedo (reflectivity) of the land and sea surfaces that are currently covered in bright ice and snow, another is the release of a carbon that is stored in ice, and a third is the rise of sea level that is resulting from the melting of glacial ice on land.

The albedo effect of melting glacial ice and floating sea ice is a rapid and strongly positive climate feedback, one that did not exist during the Paleocene and Eocene. As shown in figure 2.5, ice covered with snow has an albedo of between 70% and 90%. The bare ground that is exposed when a glacier melts has an albedo close to 30%, while the open water exposed when sea ice melts has an albedo of less than 10%. A 2014 study showed a 3% decrease in the albedo of the Arctic Ocean region because of melting sea ice between 1980 to 2010.[2] The authors concluded that the global warming that has resulted solely from this albedo change is equivalent to approximately one-quarter of the warming that is attributed to increased global CO_2 levels over the same period. That's only the Arctic region, and it's just sea ice (not glacial ice on land), so it's not difficult to see that the albedo change associated with melting ice is a significant one, and it's a factor that did not exist during the PETM. The ice-albedo feedback could make our glaciated world warm much faster than the unglaciated world of the early Eocene.

As shown in figure 7.1, the rate of Arctic sea-ice loss is alarmingly fast. In just 42 years, between 1980 and 2022, the volume of September

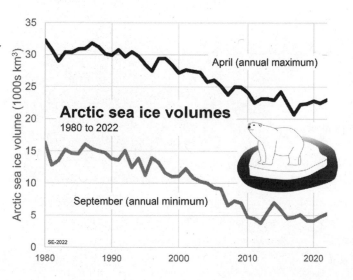

Figure 7.1: *Arctic September minimum and April maximum volumes from 1980 to 2022. (Based on Polar Science Center data.[3])*

ice decreased by 70%, while the volume of April ice decreased by 30%. (The minimum Arctic sea-ice cover is in September of each year; the maximum is in April.[4]) Based on these numbers, it is projected that the Arctic Ocean might be completely free of ice in September as early as 2035. The complete (year-round) loss of Arctic sea ice will take longer, but it is likely to be essentially gone by the end of this century.

The extent of sea ice around Antarctica is similar to that in the Arctic Ocean, but its area has not been decreasing as quickly as in the Arctic—mostly because the Antarctic region has not warmed as fast. In fact, there was a slow increase in Antarctic sea-ice extent from 1982 to 2014. That turned around in 2015,[5] and although we don't yet know if the turnaround is just a blip, another record summertime low was set in February of 2023,[6] and an alarmingly strong record for wintertime low was set in September of the same year.[7] There is little doubt that most of the Antarctic sea ice will eventually melt as well. That melting will have an even greater ice-albedo feedback effect because, as shown in figure 7.2, the sea ice surrounding Antarctica is at generally lower latitudes than that in the Arctic. That means that there is more intense sun shining on the Antarctic sea ice, so when that melts, the extra warming effect will be greater than for the Arctic.

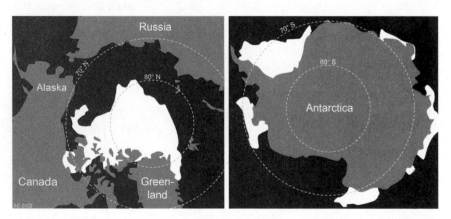

Figure 7.2: *Typical minimum sea-ice extent in the Arctic (left) and the Antarctic (right) at the same scale. Most of the Arctic sea ice is north of 80° N, while none of the Antarctic sea ice is south of 80° S. (Based on maps created by the National Snow and Ice Data Centre available at nsidc.org)*

We can count on between 1 and 2 meters of sea-level rise by the end of this century, and that means that a lot of places where people now live will be under water. Of course, this is just the beginning of anthropogenic sea-level rise. The IPCC reports that sea level is committed to rise for centuries because of the carbon that we have already emitted. It will reach between 2 and 6 meters if warming is limited to 2°C, and between 19 and 22 meters if warming is limited to 5°C.[8] If we reach the magnitude of warming that was experienced during the PETM (around 7°C), almost all glacial ice will eventually melt (although that would take thousands of years), and sea-level rise will exceed 65 meters. That didn't happen during the PETM because there was no glacial ice, but even if it had, the consequences would not have been dire because most organisms could have moved inland. Our problem is that much of our infrastructure and farmland is situated in areas that are less than 10 meters above current sea level.

Sea-level rise also contributes to an overall change in global albedo because water is less reflective than the land that is submerged by rising seas. The percentage of land that will be flooded by a 65-meter rise in sea level is quite small—less than 1%—but the albedo impact will still be significant because much of that land is in tropical or near-tropical regions. Albedo feedbacks are much more pronounced in low-latitude parts of the Earth that have stronger sunshine.

Sea-ice loss is a very fast and effective albedo feedback. Land ice (i.e., glacial ice) albedo feedback is also significant. Most mountain glaciers are currently receding at alarming rates. The authors of a recent study predict that all the glacial ice of central Europe will be gone by 2100, as will 99% of the ice in western North America, 96% of the ice in Scandinavia, and 62% of the large area of ice in Iceland (figure 7.3). The Greenland and Antarctic ice caps are also melting quickly, but they are large enough that it will take much longer (thousands of years) for their shrinkage to have a significant albedo effect.

Greenhouse gas feedbacks are another matter, and there are several of those. The first to bite us will be destabilization of permafrost, which is also already underway.[9] That, and the related destabilization and then erosion of peat deposits and soil because of terrain instability and slope

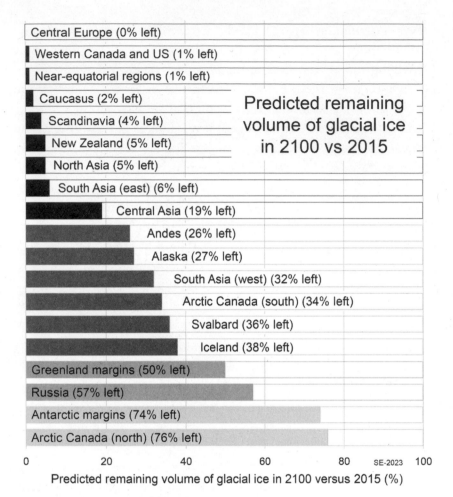

Figure 7.3: *Predicted percent area of remaining glacial ice in 2100 versus 2015, based on the IPCC's "business as usual" scenario.*[10] *(The main Greenland and Antarctic ice sheets are excluded.)*

failure, could eventually release over 2,000 billion tonnes of carbon, pushing us well into PETM territory. Permafrost deposits are currently warming by about 0.3°C per decade[11] and so, over the next century, will likely warm by a total of 3°C, or probably more because the rate of warming is accelerating. Since roughly half of permafrost deposits are already warmer than -3°C, a significant proportion of permafrost is at risk of melting within the lifetimes of our children.

The other looming carbon reservoir is that of seafloor methane hydrates, and while the deep-water deposits are likely safe for at least centuries, the shallower ones, on continental shelves and the ones on land in polar regions, are more vulnerable.

Glacial melting also changes ocean currents. As we saw in chapter 2, the strength of surface and deep-water thermohaline circulation in the Atlantic Ocean is controlled by density related to salinity. If the salinity decreases because of fresh water from rapid melting of Arctic glaciers, the deep circulation slows, with repercussions around the world, such as deep-ocean anoxia, deep-ocean warming, and a reduction in the rate of transfer of CO_2 from the atmosphere to the oceans. A recent study of the Southern Ocean has shown that a similar reduction in deep circulation there can be ascribed to accelerated glacial melting of the Antarctic ice cap.[12] Matt England, one of the authors stated: "In the past, these circulations have taken more than 1,000 years or so to change, but this is happening over just a few decades. It's way faster than we thought these circulations could slow down."[13] This change could not have happened as quickly in the late Paleocene because there was no glacial ice to melt.

There is carbon stored in both sea ice and glacial ice, and although the concentrations are relatively low compared with other carbon reservoirs, the sheer volume of ice, and the potential loss of most of the Earth's glacial ice in the coming millennia, makes it a factor. It is estimated that glaciers contain about 6 billion tonnes of organic carbon.[14] That is a small amount compared with the hundreds to thousands of billions of tonnes in other reservoirs, but it will contribute to the warming feedback process.

Humans

The human population passed the eight billion threshold in November 2022, and while the rate of growth is slowing, we will exceed nine billion before our numbers start to decline. The total biomass of humans plus our domestic livestock (cows, sheep, chickens, etc.) is almost 18 times that of all other terrestrial vertebrates combined (i.e., wild mammals, reptiles, birds, etc.), so it is no surprise that we have changed the face of the Earth. We have initiated the Anthropocene by replacing forests with cattle pasture and wild grasses with salad greens, by exterminating entire species and decimating many others, by draining swamps and damming rivers, and by burning massive amounts of fossil fuel, much of it to make stuff we don't need and to travel to places that we don't need to visit.

The most significant of these actions is our conversion of the carbon stored in fossil fuels into CO_2, which, as shown in figure 7.4, reached a cumulative total of 474 billion tonnes by the end of 2021 (and will almost certainly surpass 500 billion tonnes early in 2024). Compare that with the total reservoir of fossils fuels of approximately 2,500 billion tonnes, and with some of the other reservoirs such as permafrost (~1,700) and biomass (~500), as illustrated in figure 6.1.

Figure 7.4 shows that CO^2 emission rates have likely not yet peaked. If we adhere to the 2015 Paris Agreement by making sensible individual decisions and convincing our governments to enact increasingly strong national policies, our emissions should peak within a few years, and we could find ourselves on the "Aspirations to stay below 2°C" curve. Alas, the shaded area under that curve is quite similar to the area of our historical emissions, so even in this best-case scenario, we will likely end up emitting a total of close to 1,000 billion tonnes of fossil

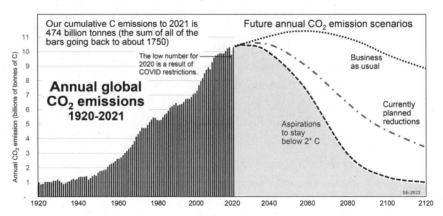

Figure 7.4: *Cumulative global CO_2 emissions from 1990 to 2021 and some approximate potential future emission scenarios. The graph shows the annual amounts of CO_2 (expressed as C) emitted as a result of human activities from 1920 to the end of 2021. It does not include carbon emissions from methane or from land-use changes.*[15]

fuel carbon before we're done. If we change virtually nothing ("business as usual"), our use won't peak for several more decades, and our total fossil fuel carbon emissions will likely exceed 2,000 billion tonnes.

The other main anthropogenic GHG, methane, is responsible for about 32% of present-day climate warming. Most of our methane emissions result from our craving for beef and dairy, and from carelessness in the production and processing of fossil fuels.

Some 10,000 years ago, before humans started significantly changing their environment, about 57% of the Earth's useable land (excludes very steep slopes, bare rock, glacial ice, and desert) was forested, and the remaining 43% was natural grassland. As illustrated in figure 7.5, that changed very little for the following 5,000 years, with only a very small area (about 1%) taken up by crops. By 1700 AD, about 10% of useable land was farmed, and by 2018, the area used for urbanization, crops, and especially for grazing (mostly for beef cattle) had increased to 48%. Natural grassland is now down to only about 16%, while forests comprise 38% of useable land. And many of those are not the diverse and healthy old-growth forests that they used to be. Instead, they are second- or third-generation monoculture tree farms, many of which are surviving on chemical fertilizers and riddled with logging roads and yahoos on all-terrain vehicles.

The important point of figure 7.5 is that we have replaced about 48% of the area of Earth's natural vegetation with pastures, farms, and cities. That has resulted in the release of sequestered carbon (from vegetation and soil) and a significant reduction of the land's ability to sequester more carbon.

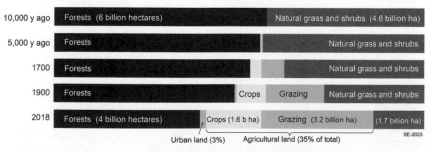

Figure 7.5: *Global changes in land cover over the past 10,000 years.*[16]

The Earth of the late Paleocene likely had a forest/grassland ratio similar to that of 10,000 years ago, although the ecosystems were adapted to a warmer climate than ours. We know that as the temperature increased in the early Eocene, forests changed in many areas, and it's possible that a lot of biomass carbon was released in the process, adding to the positive feedback. Today's compromised forests would probably suffer even more, but because of what we can see in figure 7.5, there is less forest to give up its carbon. Relatively little carbon is stored in cropland or grazing land.

Other Differences

One of the important differences between now and the Paleocene is that the movement of continents has resulted in changes to some key ocean currents. In the Paleocene, the Antarctic Peninsula was close enough to the southern tip of South America to prevent the significant flow of water between the two continents. By about 20 million years later (around 35 Ma, late Eocene), the two land masses had moved sufficiently apart for the development of a strong east-flowing current through the Drake Passage and around Antarctica (figure 7.6). That current effectively

Figure 7.6:
Approximate path of the Antarctic Circumpolar Current.

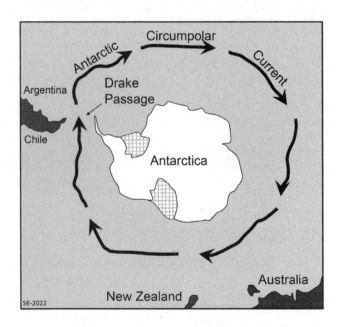

separated Antarctica from the warmer parts of the Pacific, Atlantic, and Indian Oceans, and contributed to the growth, on Antarctica, of the Earth's first significant glaciers for over 200 million years.

Another 20 million years later, volcanism built the Isthmus of Panama, blocking the flow of water through the Central American Seaway. The resulting changes in Atlantic currents eventually led to northern hemisphere glaciation.

Volcanism is happening all the time on Earth. There is usually a visible eruption taking place on land somewhere, and if there isn't, there is almost certainly one on the seafloor. But the steady rhythm of small, medium, and even large eruptions doesn't change the climate very much, or at least not for very long. An episode of higher-than-average volcanism that lasts tens of thousands of years can warm the climate because of the gradual buildup of CO_2. The strong volcanism that may have caused the changes that led to the most severe extinction on record—at the end of the Permian (250 million years ago)—is thought to have lasted a total of about 400,000 years.[17] The less intense North Atlantic Igneous Province volcanism, some small part of which might have triggered the PETM (see chapter 6), lasted for several million years.

But no extraordinary volcanism is happening now, just some infrequent explosive eruptions like Pinatubo in 1991, or the Hunga Tonga underwater eruption early in 2022, and some small off-and-on lava flows in Hawaii and elsewhere. And we have no evidence to suggest that we're about to enter a new volcanic age, so while volcanism may have been significant in triggering the PETM, there is no reason to think that it will be a factor in our near future.

Summary

Chapter 7 provides an assessment of how present-day Earth differs from that of the late Paleocene and describes how those differences can help us to understand if we could be in for a devastating PETM-like runaway climate crisis in the near future, on top of the misery that we can expect from ongoing anthropogenic climate change.

The present day is similar to the late Paleocene in terms of the general distribution of the continents and oceans and in the existence of some large reservoirs of sequestered carbon, such as permafrost and methane hydrates that have the potential to send us over a runaway climate cliff.

Our time is different from the Paleocene because there is a great deal of ice on the Earth now, both in glaciers on land and floating on the polar seas. That ice is a climate change bomb because it has the potential to significantly amplify warming through albedo feedback and changes to ocean currents. It will also contribute to sea-level rise that will displace billions of people and their infrastructure.

The other main difference is many billions of humans, and our reckless use of fossil fuels and our craving for beef and dairy—through which we are adding carbon to the atmosphere at a rate which is likely at least five times faster than that during the early part of the PETM—and our massive changes to the landscape. Another important difference is that there is only a low level of volcanism at present, not enough to change the climate.

Part II
Where We Are Heading
If We Don't Change Course

Chapters 8, 9, and 10 include descriptions of some of the potential outcomes of ongoing climate change compared with the more serious outcomes of a PETM-like runaway climate. People who write or speak about the climate are often advised to remain upbeat, and to avoid the risk of scaring or depressing readers into despair and consequent inaction. That may be good advice, but it can also be argued that it's important to be realistic because we are responsible adults. All of us have contributed to climate change, and we need to be aware of what could be coming, not just for us, but for our children and grandchildren.

So here's the warning: if we don't get our shit together quickly (within a decade), our lives, and especially those of our descendants (including my own grandchildren), could get extremely difficult. Furthermore, if the climate transitions to a PETM-like state, the consequences will be many times worse. If you don't want to read about that, then I reluctantly advise you to skip to chapter 11. That said, I do believe we can turn things around—that we really can avoid a catastrophic climate scenario—otherwise I wouldn't bother writing this.

Chapter 11 provides a summary of what we need to do to get there. It will be hard work, but there will be many positive side effects.

Chapter 8

How the Oceans Might Change
Under PETM Conditions

An unprejudiced observer might well be fearful that in the not too distant future our children may be able to learn about the coral reefs only from books and documentary films, for one of nature's most unique habitats will have vanished from the face of the Earth.

—Gilbert Voss

As DESCRIBED IN CHAPTER 4, the oceans play a key role in the Earth's climate system; what happens within and above them has profound implications for the conditions on land and our ability to live here. We know that there were dramatic changes to the oceans during the PETM, and we can expect that similar things will happen if a PETM-like runaway climate takes hold in the near future. The goal in this chapter is to understand the potential physical changes in the oceans and how they will affect us, both directly through altered weather systems and indirectly through the marine ecosystems that we depend on.

The following projections are based on observations of the effects of anthropogenic climate change to date, on the predictions of climate scientists for anthropogenic climate change over the rest of this century, and on the interpretations of what happened during the PETM based on the geological record. Evidence suggests that it took around 1,000 years for the PETM runaway warming to take effect, but as discussed, there are good reasons to think that a future runaway could be much faster. That's because of the dangerously high rate of anthropogenic GHG emissions and the very fast feedbacks that we'll get from melting sea ice and glacial ice. If we don't act decisively over the next decade, it is quite possible that PETM-like conditions could start to take hold on Earth early in the next century, or even sooner. People born within the past decade—including my own grandchildren—could face that reality.

Ocean Water Temperature

We have already seen an increase of global average sea-surface temperatures of just under 1°C. According to the IPCC *Sixth Assessment Report* (2021), it is likely that sea surface temperatures will increase to 2° to 3°C above pre-industrial levels by the end of this century entirely due to expected anthropogenic warming, with even greater increases (more than 4°C) in some Arctic regions.[1] The deeper ocean is also already warming, but that process will be much slower than surface warming.

During the PETM, the ocean surface waters warmed by up to 8°C (figure 4.1), and eventually strong warming extended to even the deepest waters. That amount of ocean warming had significant implications for major storm systems—such as tropical cyclones—that get their energy from warm water, and for increased intensity of rainfall, reduced deeper-ocean oxygen levels, destruction of coral reefs, and significantly reduced productivity of marine ecosystems.

Even with just a single degree of surface ocean warming, tropical storm numbers, intensity, duration, and destructiveness have increased. Over the remainder of this century, the proportion of cyclones that reach category 4 and 5 is expected to be greater still, and there will be much more rainfall on land.[2] Under a PETM scenario, even stronger tropical storms will contribute to more severe destruction of near-coast regions than already seen or even forecast. There will also be severe inland flooding from tropical storm rainfall, including mega-storm events that recontour the terrain over huge areas, as happened in southern Europe during the PETM (chapter 5).

The deeper parts of the oceans are already becoming depleted in oxygen as a result of anthropogenic climate change, firstly because warming is slowing deep-ocean circulation patterns and secondly because oxygen is less soluble in warm water than in cold water. This effect is expected to increase in coming decades, and even more under a PETM scenario.

These lower oxygen levels in the moderate and deep parts of the ocean will have implications for all ocean life.

Coral reef communities are particularly vulnerable to ocean warming, and reef mortality has been happening over large areas during very warm years since the 1980s. The projections for reef health over the next several decades are dire; 50% of reefs are already seriously bleached and degraded (figure 8.1), and it is expected that 90% will be lost by 2050.[3] Loss of reefs will seriously reduce habitat for a huge range of marine organisms, will contribute to marine oxygen depletion, and will result in much greater damage to shorelines by storm events. As we saw in chapter 4, reef communities were extirpated from a vast area during the PETM; we could expect that to be repeated in a future PETM-like world.

Warming water has overall negative impacts on ocean productivity for two main reasons: many species do not thrive in warm water, and warm water holds less oxygen than cold. It is projected that total marine animal biomass will decline between 15% and 30% due to warming in

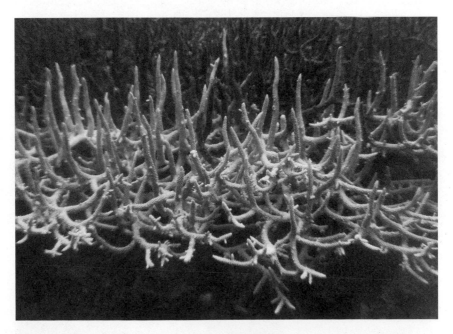

Figure 8.1: *Bleached staghorn coral* (Acropora cervicornis) *near to Cairns, Australia. (Photo by Matt Kieffer.[4])*

all tropical and temperate ocean bodies over the next 75 years; PETM conditions will make this much worse, although the smaller areas of polar ocean might see biomass increases during that time.[5]

Ocean Water Acidity

An increased atmospheric CO_2 level leads to increased acidity of the oceans because a large proportion of atmospheric CO_2 gets dissolved in the ocean and is converted to carbonic acid. That phenomenon has been observed over the past several decades and is projected to increase in the coming decades. This is consistent with the PETM, when there was a pronounced shallowing—by over 2,000 meters—of the depth at which calcium carbonate became soluble in the ocean because of increased acidity (figure 4.5).

All marine organisms that make hard parts of calcium carbonate will be affected by its increased solubility as acidity increases. Small organisms with thin shells are likely to be affected first, especially those living at high latitudes. One example is the sea butterfly (*Limacina* sp.) which lives in both north and south polar oceans. *Limacina* has a thin, nearly transparent shell (figure 8.2) and has been shown to build significantly thinner and weaker shells under conditions that approximate the acidity of near future oceans.[6]

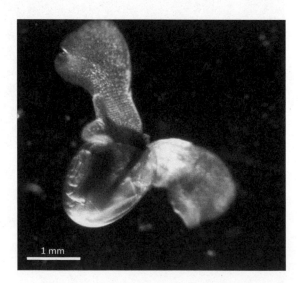

Figure 8.2: *The arctic sea butterfly,* Limacina helicina. *(Photo by Ross Hopcroft.[7])*

Further increased acidity will affect tens of thousands of calcium carbonate-dependant marine species—some of them too small to see. These species are a critical part of the ocean's food web.

Sea-level Rise

Warming ocean water and melting glaciers both contribute to sea-level rise. We have already seen about 20 centimeters of global sea-level rise, and the rate of increase is also increasing. We can expect sea level to be higher by between 100 and 200 centimeters by the end of this century. The problem with future and PETM-scenario sea-level rise is that there are likely to be episodes of very rapid sea-level rise due to the episodic collapse of Greenland and Antarctica ice sheets. We are currently witnessing vast areas of floating ice break off from Antarctica and Greenland, and while those don't contribute to sea-level rise (because the ice is already in the water), they make the large glaciers behind them more vulnerable to rapid seaward advance, and therefore to the potential for sea-level rise many times faster than we are currently observing. An example of that is the vulnerable Thwaites Glacier in Antarctica, which is being temporarily held back by floating ice shelves and under-ice topography.[8]

Ocean Currents

Ocean currents, including the well-known surface currents like the Gulf Stream and the less well-known deep-ocean currents like North Atlantic Deep Water (NADW), are fundamental to the health of the oceans and ocean organisms, and also to the maintenance of the Earth's equable climate, meaning a climate that is not overwhelmingly hot in equatorial regions. The evidence for slowing and possible imminent (within decades) collapse of the Atlantic Ocean circulation patterns (including the Gulf Stream and the NADW) is growing.[9] A major change to this global circulation system will result in relative cooling of the North Atlantic (and neighboring land areas) and intensified heating in the southern hemisphere, and would likely also disrupt precipitation patterns, putting billions of people at even greater risk than they are now. There is also clear evidence of a strong decline in the deep-ocean current that originates in the Southern Ocean.[10] Changes to these currents will have dramatic effects on our climate, including cooling of some temperate regions (e.g., northwestern Europe), heating of tropical regions, and disruption of the monsoon patterns that bring life-giving water to billions of people.

But these observed and predicted changes to the ocean currents are nothing compared to what is likely to happen if we slide into a PETM-like climate. As described in chapter 4, ocean circulation patterns were completely rearranged during the PETM. Formation of deep currents, which had been taking place in polar regions because of the high density of cold water, shifted to equatorial regions where strong evaporation made the seawater very salty, and therefore dense enough to sink and initiate deep currents. Unlike the cold polar currents, those deep currents helped to bring warm water to the ocean depths and likely contributed to the demise of many deep-water microorganisms and everything that depended on them. It is also evident that, eventually, these PETM deeper currents did not reach the depths that they had in the Paleocene, or do today.

Changes in deep-ocean currents would necessarily be accompanied by changes to surface currents, and while it's hard to predict what those changes might look like, there is no doubt that they would have had implications for major weather systems like monsoons and jet stream trajectories, and also for local weather patterns.

Implications for Humans

In June 2021, an intense heat dome, an air mass trapped by an anomalous loop in the jet stream, established itself over the Pacific Northwest region of North America, especially in Washington and British Columbia. Temperatures exceeded 40°C for several days (see box 10.1). It was estimated that over a billion shellfish were literally cooked in their shells on rocky intertidal zones.[11] The smell of death was intense. Two years later, the intertidal zone appears to have recovered, but this is a tiny taste of how ocean ecosystems, including those that humans depend on for food, will be affected by rising temperatures related to anthropogenic climate change. PETM-like climate change would be many times worse.

If we don't quickly reduce our emissions and also take other steps, such as reforestation, to bring the climate under control, there is a very real risk that feedback processes could take over and flip the Earth into a PETM-like runaway climate state. This could start to happen before the end of this century. The likely consequences of that scenario for the oceans are summarized below and in figure 8.3.

Some of the implications for humans would be as follows:

- The oceans will get warmer everywhere. That will lead to dramatic declines in marine productivity at all levels in the food web, including those that humans depend on for food. Most species will migrate to higher latitudes, forcing food fisheries to change their strategies or move along with them.
- High water temperatures will result in the complete extirpation of coral reef communities from tropical regions, leading to a further decline in ocean productivity and ocean water oxygen levels. There will be reduced seafood resources for humans. There will also be reduced storm protection for near-shore communities

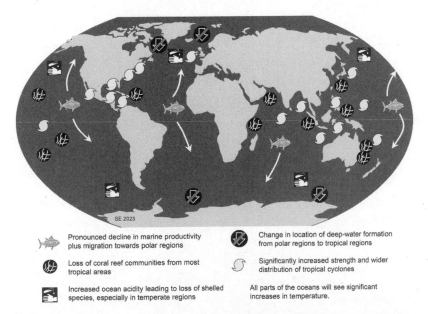

Pronounced decline in marine productivity plus migration towards polar regions

Loss of coral reef communities from most tropical areas

Increased ocean acidity leading to loss of shelled species, especially in temperate regions

Change in location of deep-water formation from polar regions to tropical regions

Significantly increased strength and wider distribution of tropical cyclones

All parts of the oceans will see significant increases in temperature.

Figure 8.3: *Summary of the ocean impacts of a near-future PETM-like runaway climate.*

that are already trying to cope with sea-level rise. It is possible that reef organisms will find localized refugia in more northern or southern waters.

- The oceans will become increasingly acidic. This will also affect reef communities, but the most significant impact will be on very small floating and swimming organisms that make their hard parts out of calcium carbonate. The decimation of these species will impact the entire ocean food web, and so will affect human food supplies.
- There will be a dramatic change in the patterns of ocean circulation, including where, and the extent to which, surface ocean water sinks to depth. These changes will also create negative outcomes for marine ecosystems, and those that depend on them, and could significantly change monsoon behavior, rendering some areas that are now heavily populated uninhabitable because of repeated flooding or long-term drought.

Box 8.1 Sea-level Rise and Beaches

If you're anything like me, you probably like spending time relaxing or cooling off at the beach. Perhaps you like jetting to exotic places with sunny beaches and tropical cocktails, or maybe there's a nice beach near enough to get to on hot summer weekends.

If your favourite beach is anything like the one in this photo[12]— Copacabana Beach in Rio—it is at serious risk from climate change. Some beaches are backed by rocky cliffs. They'll have nowhere to go and will simply be submerged as the water rises. This beach is hard up against a busy road and a massive residential and commercial zone. Over the next few decades, Copacabana Beach will get narrower and narrower, and water will wash over that road during storms. Eventually the Atlantic Ocean will spill into the streets of Copacabana.

Beaches that are situated near large rivers that bring lots of sand to the coast, or in locations with an ongoing supply of sand behind the beach, may last longer, but before too long, almost all of the world's beautiful

- The warmer oceans will generate more intense tropical storms, and fueled by warm water at higher latitudes, those storms will almost certainly extend into regions that do not typically experience tropical storms today, such as Europe, the northeastern US and Canadian Atlantic provinces, California, New Zealand, and southern Australia. Coastal communities, already more vulnerable because of higher sea levels, will be devastated and the flooding from torrential rains will reach far inland.
- Sea-level rise is not shown on figure 8.3, but the implications of sea-level rise for humans are obvious, with more areas that are currently highly populated rendered uninhabitable, forcing more people to move. Sea-level rise is currently around 4 centimeters per decade. The effects of sea-level rise on land areas are discussed in more detail in chapter 9.

beaches will disappear because the geologically slow process of beach building won't be able to keep up with the accelerating pace of sea-level rise. For example, the authors of a recent study stated that "by 2100, the model estimates that 25 to 70% of California's beaches may become completely eroded due to sea-level rise scenarios of 0.5 to 3.0 m, respectively."[13]

Summary

There is a credible risk that PETM-like runaway climate change could start to take hold in the near future, possibly as early as the end of this century. That would bring significant changes to the oceans, including strong warming of near-surface water and eventually deep water, increased acidity, wholesale rearrangement of both surface and deep-ocean currents, increased intensity and extent of tropical storms, devastation of coral reef ecosystems, and decimation of many other marine ecosystems. All these changes would have severe implications for human food supplies and for the suitability of coastal regions—and even some inland regions—for habitation. Billions of people would be forced to migrate because of changes in the oceans.

Chapter 9

How the Land Might Change Under PETM Conditions

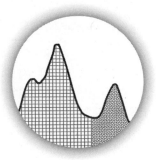

> *He sat up. The air was still hotter than the water. He watched*
> *the sunlight strike the trees on the other side of the lake; it looked*
> *like they were bursting into flame. Balancing his head carefully*
> *on his spine he surveyed the scene. Everyone was dead.*
>
> —Kim Stanley Robinson,
> *The Ministry for the Future*

I LIVE IN A TEMPERATE RAIN FOREST, a region where giant 1,000-year-old conifers were once common; most of the big ones have been cut down now. It is normally a cool and rainy region, and it is not one of the places on Earth that has been particularly brutalized by climate change, at least not so far. And yet as I look around me, even here, I see troubling signs. Entire groves of western red cedar are dead, and grand fir stands have been decimated. Just last month (May 2023), dozens of all-time daily temperature records were broken, day after day. A campfire ban (which normally only happens in the summer) was in place before the end of May, and the mountain I watch to gauge the snowpack was nearly bare in early June, whereas it used to have an extensive snowpack well into September.

Although the oceans are critical to our survival on Earth, we are land animals, and so what happens on the land is what affects us the most. As in chapter 8, we will look at the changes that have been observed to date, what is predicted through the standard climate change models over the next several decades, and what could happen if we let the Earth change enough so that runaway feedbacks take us into a PETM-like state, and how that would likely affect us.

Temperature

As shown in figure 9.1, the oceans have warmed by about 0.6°C since the reference period of 1951 to 1980, but the change on land has been significantly greater, closer to 1.4°C. IPCC projections for global temperature increases (land and ocean areas combined) to the end of this century range from a low of about 2°C (for low-emission scenarios) to a high of around 4°C (for high-emission scenarios).[1] But as we can see from figure 9.1, the increases on land are likely to be greater than those global estimates.

Geological records indicate that there was warming of 7° to 8°C in most land regions during the PETM, with the greatest increases at high latitudes (in some regions over 8°C), and the least (5° to 6°C) in tropical regions (figure 5.2).

High temperatures also play a critical role in enhancing aridity through evaporation, and so contribute water stress for all types of organisms, and to the risk of wildfires.

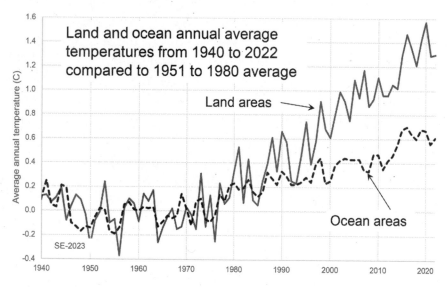

By the author using public domain data from NASA at
https://data.giss.nasa.gov/gistemp/graphs_v4/

Figure 9.1: *Record of average annual temperature change from 1940 to 2022 for land areas and for ocean areas, excluding ice-covered ocean.*[2]

Precipitation and Aridity

Warm air can hold more water than cold air, and so the atmospheric warming to date has made for a wetter atmosphere, and for generally higher precipitation overall. Predicted patterns of precipitation change to the end of this century are summarized in figure 9.2. To some degree, areas that are now dry will get less precipitation (e.g., US Southwest and Mexico, Southern Africa, Australia, Mediterranean region), while high-latitude regions (northern Canada, northern Asia, Antarctica) will get more. But since warm conditions lead to stronger evaporation, aridity is becoming more of a problem everywhere, even in areas that are expected to get more precipitation. Based on a study conducted in 2012, severe and widespread droughts are predicted for many regions in the coming decades as a result of decreased precipitation and/or increased evaporation.[3] Strong overall drying is also evident during the PETM based on rock characteristics at many locations.

Increased aridity leads to increases in the risk of wildfires. As shown in figure 9.3, there has been a strong global increase in forested area lost to wildfires over the past two decades. This trend is particularly worrisome because wildfires are a strong positive climate feedback since they convert sequestered carbon (in plant tissues) to carbon dioxide. Wildfires

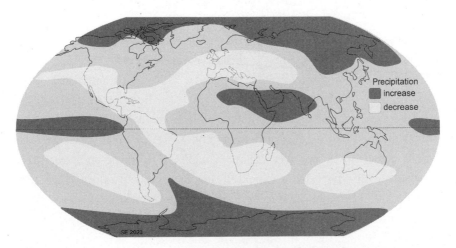

Figure 9.2: *Generalized changes in precipitation projected for the remainder of this century.*[4]

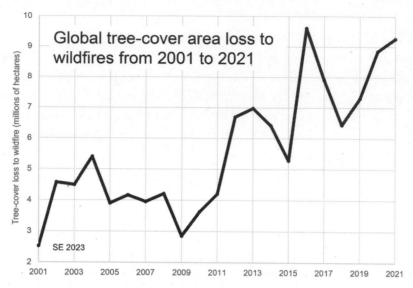

Figure 9.3: *Trend of tree-cover area lost to wildfires from 2001 to 2021. For comparison, Ireland or Austria are each around 8 million hectares.[5]*

also contribute to instability of slopes and to soil erosion, leading to even more release of sequestered carbon. There is persistent evidence that the forests that we've become familiar with might never recover in many regions affected by wildfires, especially repeated fires over periods of decades.[6]

Notwithstanding the expectation of overall increasing aridity, one of the consistent predictions of climate modelling is an increase in the frequency of intense precipitation events. Over the next century, the probability for intense rainfall is expected to increase in virtually all land areas, with the greatest increases at high latitudes and across a belt from the Horn of Africa through the Arabian Peninsula, Pakistan, India, Bangladesh, and Myanmar.[7] The 2022 flooding in Pakistan is an example of what can be expected due to increased rainfall intensity. Large areas received over 2 meters of rain between June and August of 2022, and the resulting floods left over 1,700 dead, 2 million home-less, and over 8 million displaced. The Pakistan flooding was the result of a particularly strong South Asian monsoon season. Other unusual flooding events, for example those in western North America in the past few years, have been called atmospheric rivers, where streams of extraordinarily moist air from the Pacific Ocean bring intense rains for a few days at a time. Flooding is also a common result of landfalling tropical storms, and as those are likely to become more powerful in the coming decades, and even more powerful under a PETM scenario, this kind of flooding is something that we can expect to see more frequently and with greater severity. The PETM-aged Claret Conglomerate of northern Spain provides evidence of flooding and very rapid water flow on a scale unlike anything that has been observed in historical times.

Increasing temperatures over the past few decades have resulted in diminished snowfall accumulations in temperate and mountainous regions. For example, North America experienced an approximate 12% decline in snow accumulation from 1980 to 2020.[8] This trend is projected to continue as temperatures increase over the next several decades. Furthermore, as described in chapter 7 (see figure 7.1), alpine glaciers and some of the smaller continental glaciers are expected to be mostly gone by the end of this century. Under PETM conditions, all glaciers except the Greenland and Antarctic ice sheets would disappear, and those would be melting rapidly. Snowfall would be infrequent and minimal except at very high latitudes and elevations.

Changes to Ecosystems

Almost everyone who has lived in one place for a few decades has at least one story about changes in their ecosystems, and many people have the data to back up their stories. Ecological responses to climate change documented by scientists over the past several decades are summarized in a highly referenced 2006 review by Camille Parmesan[9] as follows:

- the advance of springtime events such as flowering, migrating, and breeding—in some cases by several weeks—has been observed globally;
- changes in the timing of behaviors and migration of interacting species have resulted in asynchrony in predator-prey and insect-plant relationships, in most cases, with negative consequences, such as pollination failures;

- poleward range shifts of both plants and animals in the order of tens of kilometers to well over 100 kilometers over just decades have been documented on all continents;
- range-restricted species (on mountains or in polar regions) have experienced range contractions and extinctions; and
- shifts in the ranges and abundances of parasites and parasite vectors are influencing human health.

There have been many more such changes since 2006, and there will be increasingly more over the coming decades. The IPCC has projected that 9% to 14% of species assessed will be at very high risk of extinction with 1.5°C of warming. That rate is predicted to climb to 12% to 29% at 3°C and to 15% to 48% at 5°C.[10]

As noted in chapters 1, 2, and 5, there were massive changes to terrestrial ecosystems during the PETM. Conifers were extirpated from many areas and were replaced by broadleaf species that had formerly lived hundreds to thousands of kilometers closer to the equator. Vertebrates were also dramatically affected by the changed climate, with range shifts of thousands of kilometers for some. Although there were some extirpations from the areas studied, relatively few extinctions were recorded (although there may have been many more that were not recorded), and several new groups evolved. Ecosystems were forced to adapt not only to significantly increased temperatures but also to reduced availability of water.

Implications for Humans

The Earth's mean annual temperature (MAT) in both land and sea areas is currently about 15°C, which is up from 14°C just a few decades ago. It is on its way to 16°C within another few decades. By that time, the Earth will be warmer than it has ever been over the 2.8 million years that *Homo* has existed, and large parts of it will simply be too hot for human survival.

Figure 9.4 shows the range of MATs at which most humans live. There is a primary peak at around 15°C and secondary one at around 26°C. The primary peak represents the ideal niche for human survival in relatively dry climates, while the secondary peak represents a niche of survivability in areas where there are strong and reliable summer rains (e.g., monsoons) or reliable river flows. As shown on the diagram, China, Korea, Japan, the US, and most European countries are within the primary niche. Russia and Canada lie off to the cool side of this niche, but in fact almost all Canadians and Russians live in the warmer parts of their countries, actually within the primary niche. These are the A and B countries on the diagram. Many other A and B countries with smaller populations are not listed. Another group lies on the warmer side of the primary niche, including Uganda, South Africa, Morocco, Iran, Iraq, Pakistan, Mexico, and Argentina (and many other less populus countries). These are the C countries on the diagram. Many Asian, African, and South American countries fall within the secondary warmer and wetter niche, including India, Indonesia, Bangladesh, Nigeria, Kenya, Egypt (because of the Nile), and Brazil; these are the D countries.

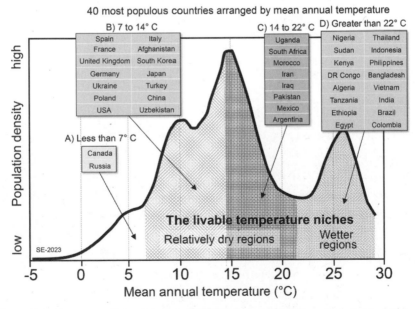

Figure 9.4: *Mean annual temperatures (MAT) that are suitable for human survival. The heavy line is from Lenton, and it represents the actual distribution of humans in 2010 based on their local MAT. The countries listed are grouped based on their national MATs (local MAT can vary quite widely from the national MAT).[11]*

It is difficult to survive in areas with mean annual temperatures greater than 28°C, and in dry places, the limit is considerably lower, unless most of the food is imported. Indeed, no country currently has a national MAT above 28°C, although many uninhabited or sparsely populated places within the warmer countries do. As the global temperature increases, everyone is going to get pushed to the right-hand side of figure 9.4. Most people in the A and B countries will still be OK (at least for now), but large regions within many of the relatively dry C countries and the hot and humid D countries will become too hot—and increasingly too arid—to be liveable. As the temperature continues to rise, most countries will be affected. If you currently live in an A or B country and you are thinking that your grandchildren are going to be OK, then think again.

In the context of elevated temperatures, it is critical to consider how survivability is influenced by humidity. The key parameter that determines our ability to survive heat is the wet-bulb temperature because high humidity in combination with high temperature limits our ability to keep cool (box 9.1). But it's not only about wet-bulb temperature.

Box 9.1 How Hot Is Too Hot?

The genus *Homo* evolved around 2.8 million years ago. Since then, the Earth has never been significantly warmer than it is now, and it has often been cooler. Owing to that climate history, there is a physiological limit to how much heat humans can take; it is determined by how well we can cool ourselves through evaporation of sweat, and the limit is controlled by the wet-bulb temperature (WBT) (see figure). A wet-bulb thermometer has a wet cloth sleeve around its bulb. Evaporation of water from that sleeve cools the bulb. High humidity reduces the rate of evaporation and so reduces the cooling effect on the wet thermometer bulb, just as it reduces the cooling effect of evaporation from our skin. At 100% humidity, there is no evaporation, and so the WBT equals the dry-bulb temperature, and humans have no evaporative cooling at all. Humidity is the key. In very hot conditions, over 40°C, for example, we can be quite comfortable as long as the air is dry. The problem arises when it is too humid for our sweat to evaporate, which means that our cooling system shuts down.

A WBT above 25°C can be fatal for the elderly and the very young. When the WBT is greater than 31°C, anyone that cannot get to a cool place or into cool water will likely die within a few hours.[12]

In June 2021, normally cool British Columbia was stuck under an intense high-pressure heat dome for several days. WBTs exceeded 25°C in many places around Vancouver and reached 27.6°C in the nearby Fraser Valley. There were 619 heat-related deaths attributed to that event.[13]

Everyone should be aware of the implications of elevated WBTs and the risks associated with excess heat because that is the road we are venturing down. The likelihood of heat crises increases with every passing year.

Kim Stanley Robinson provides a frightening illustration of a deadly WBT event in his book *The Ministry for the Future*.[14] The WBT part is in chapter 1, which is only twelve pages long. I urge you to read it. It could save your life or compel you to change it!

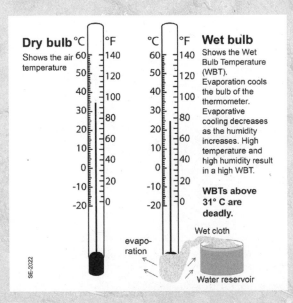

We can survive very high temperatures (over 45°C) for a short time because of the evaporative cooling, but there are limits to how much we can sweat, and how much of the resulting dehydration we can cope with.

Lenton and co-authors argue that by 2030 some 300 million people will be within areas that are hotter than the liveable limit, and that this will increase to 2.7 billion by 2090 as warming approaches 3°C above pre-industrial levels.[15] Needless to say, most of these will be people in the C and D countries of figure 9.4, people with low incomes, limited resources, and few options. It will not be easy for them to migrate nor to find places to migrate to, but if they don't leave, they will die. There will be a lot of temperature refugees, and it is going to be everyone's problem.

Figure 9.5 provides a different perspective on the livability issue. Currently some uninhabited parts of the Sahara region of Africa and a small part of the Arabian Peninsula have MATs greater than 29°C. By 2070, this 29° + uninhabitable area is expected to be many times larger, and will include part of Central America, much of northern South America, much of northern Africa, most of the Arabian Peninsula, almost all of Southeast Asia, a big part of northern Australia, and most

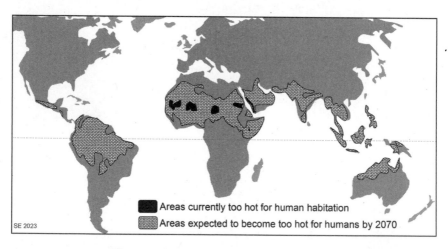

Figure 9.5: *Areas of the Earth that are currently and expected to be too hot for human habitation by 2070 under a "business as usual" climate scenario.*

of India, Pakistan, and Bangladesh.[16] These regions are currently home to about 3 billion people.

If a runaway climate starts to take hold towards the end of this century, and the temperature soars to PETM-like levels of 6, 7, and even 8°C above the pre-industrial temperature, almost all places where humans now live will become uninhabitable. Almost everyone will have to migrate to a higher latitude, and almost everyone will be a refugee.

Decreasing Snow

As noted above, warmer temperatures result in reduced accumulation of snow and earlier spring snowmelt, and this has implications that go far beyond opportunities for snow sports and scenic landscapes. In the northern hemisphere alone an estimated two billion people rely on water supplies that are supplemented by winter snow packs, and that most of those will face water shortages—both for domestic needs and for irrigation of crops—with the warming coming over the next several decades.[17] Under PETM-like conditions, we could expect snow accumulations to be minimal at best. Few will be able to depend on snowmelt to supplement their water supply, and that will make water scarcity even more of an issue.

Increasing Aridity

If reduced snow accumulation doesn't make your region difficult to live in, there is a good chance that the general increasing aridity of the planet will. As noted above, many areas will get drier with the climate change that's expected over the next several decades because the higher temperatures will result in increased evaporation. Aridity is likely to be especially acute in areas that are already on the dry side, but even many regions that are currently humid will get much drier. This will impair our ability to grow crops and raise livestock and even to keep ourselves alive, and it will get even worse if we transition to a PETM-like climate.

Lack of water could well be the most serious consequence of climate change over the next century, but, ironically, too much water won't be far behind. Under present-day conditions, weather systems that bring devastating floods (like that in Pakistan in 2022 or the flooding around Houston from Hurricane Harvey in 2017) are sufficiently infrequent for local populations to recover and rebuild. With increasing warming and an increasingly intense hydrological cycle, such events will become more frequent and more severe, and some flood-prone regions will become uninhabitable, or at least unsuitable for significant population densities.

Sea-level Rise

And then, of course, there is sea-level rise (SLR), which is already becoming a major problem in low-relief coastal areas, especially ocean island countries. The rate of rise is quite slow at present (about 4 centimeters/decade) but is widely predicted to accelerate and reach a total of between 1 and 2 meters by the end of this century. According to the World Bank, the ten countries where most people will be at risk from SLR are: China, Vietnam, Egypt, India, Indonesia, Bangladesh, Brazil, Thailand, Philippines, and Myanmar.[18] The World Bank estimates that over 90 million people will be displaced by 2 meters of SLR. Based on an updated understanding of near-coastal elevations, others think this will be closer to 190 million people with 2 meters of SLR and as much as 630 million under a high-emissions scenario with 3 to 4 meters of SLR by 2100.[19] Even if we are successful in keeping temperature increase below 2°C through this century, sea level will continue to rise well into the next century because SLR is slow to be realized, and we are already committed to much more of it, potentially several meters by 2200.[20]

The current rate of SLR is slow enough to be accommodated by modifications to infrastructure (wharves and seawalls, for example) at least in areas where coastlines are relatively steep and where the resources for reconstruction are available. The rate is expected to accelerate significantly in the coming decades, but it will likely remain slow enough that it can be accommodated through adaptation and migration, in most cases. On the other hand, it is possible that we'll experience unmanageable surges in sea-level rise due to nonlinear behavior of some of the Greenland and Antarctic outflow glaciers.

Under PETM-like runaway climate change, SLR could be much faster. During the last deglaciation, the maximum sustained rate of sea-level rise happened around 14,000 years ago, during what is known as "meltwater pulse 1A." For several centuries, the rate of SLR was over ten times faster than it is now, between 40 and 60 centimeters per decade. That rate of rise—which is also possible under a future runaway climate scenario—would quickly overwhelm coastal infrastructure in almost all regions, and would require emergency migrations, potentially involving billions of people.[21] Eventually all glacial ice will melt (although that will likely take thousands of years), and sea level will rise by 65 to 70 meters. Figure 9.6 shows some of the larger areas that would be inundated by 70 meters of SLR, although all coastal regions will be affected.

Ecosystems are expected to be dramatically affected by climate change over the next several decades, and that will result in extinctions, extirpations, evolutions, and massive migrations. While healthy ecosystems are vital to our physical and mental health, billions of people (currently more than half of the global population) live in cities, and many of those do survive (although they may not thrive) effectively isolated from healthy ecosystems. But we cannot survive, or thrive, without sufficient food, and our food production is also at serious risk from climate change.

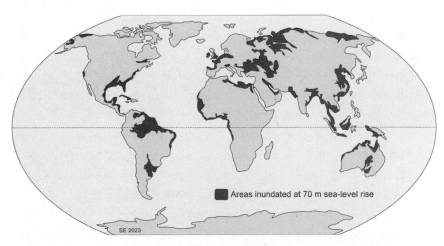

Figure 9.6: *The major areas that will be under water at 70 meters of sea-level rise.*[22]

The following table, based on a 2019 IPCC special report on climate impacts,[23] provides a summary of how food production systems are and will increasingly be affected by climate change over the next few decades, with massive negative implications for food production and quality.[24]

Table 9.1 Existing and future impacts of climate change on food production

Climate change factor	Processes	Impacts on food production
Temperature increase	Increased water demand, increased heat & drought stress, decreased growing periods, decreased soil fertility, increased pest & disease loss.	Decreased crop yields & animal performance
Elevated CO_2 level	Enhanced photosynthesis in some crops & increased water use efficiency, reduced uptake of important nutrients	Increased crop yields but decreased nutritional quality
Decreased or untimely precipitation	Drought and heat stress, crop failure, land degradation, reduced soil fertility.	Decreased crop yields, pasture stocking rates, & animal performance
Extreme events (heat, droughts, floods)	Crop failure, decrease in soil organic matter, soil erosion, disruption of distribution	Decreased crop yield, increased animal mortality

It is possible that some of the anticipated impacts on food production can be solved with future technological fixes such as genetic modifications and application of chemicals, but we need to be mindful that modern agriculture already contributes significantly to climate change (for example through deforestation, use of machinery, and the production and application of nitrogen fertilizers) and that we must significantly limit those practices to avoid making this worse. If and when we start to transition to a PETM-like climate, it is likely that the heat and aridity will become sufficiently extreme that there will be no technological fixes that could allow us to grow food in many of the places that are important for production today. Food production will have to move towards the poles. Many polar regions, such as northern Canada, which have been deglaciated for less than 10,000 years, have only thin rocky soil, so agriculture will be a challenge.

Summary

The changes to our climate that we know are coming over the next several decades will have dramatic implications for how we live on this planet. We will all be negatively affected, but there will be disproportionate challenges for people in many Global South countries that bear the least responsibility for causing the problem. What's more, if we do not take decisive action to reduce our climate impacts over the next decade, then it is possible that the changes we have caused will tip us into to a PETM-like runaway climate early in the next century. Temperatures in land areas will rise by several degrees more than we are already expecting (in a non-PETM scenario). There will be increasing drought on the one hand but increasing catastrophic flooding on the other, and it will become difficult to grow food. Some of the likely outcomes for humans are summarized in figure 9.7 and as follows:

- many regions that are currently home to billions of people will experience temperatures that are above the limits of human tolerance;
- significant reductions in snow accumulation and near complete loss of alpine glaciers will create domestic and agricultural water supply challenges for up to 2 billion people;
- increased incidence of wildfires, which is a strong positive climate feedback, will make some regions difficult to live in and will change ecosystems;
- flooding inland from super-storms and along coastlines from sea-level rise will render large areas uninhabitable, including the areas currently occupied by most of the world's major cities; and
- high temperatures, decreased or erratic rainfall, and extreme weather events will result in sharp declines in agricultural production.

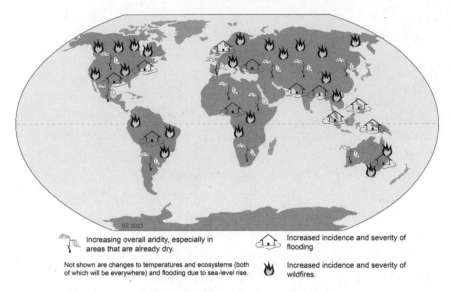

Figure 9.7: *Summary of the land impacts of a near-future PETM-like runaway climate.*

Chapter 10

Where Will Everyone Go?

The solution is so obvious it hardly needs spelling out, and yet it is rarely discussed as a serious policy: help people move for everyone's benefit.

—Gaia Vince[1]

As described in chapter 9, even the climate changes predicted to take place over the remaining decades of this century are going to put several billions of humans into situations in which they are forced to move away from where they have been living to a place where they can survive, either in their own country or in a different country. In this chapter, we will take a closer look at the origins and the consequences of the coming displacements, both for those displaced and for those lucky enough not to be displaced. As in the previous two chapters, we will consider what is happening in this context already, what is expected to happen with the kind of climate change we know is coming and we think we understand, and what could happen if we transition to a PETM-like climate.

What Is Happening Already?

Figure 10.1 shows the number of people counted by the UN High Commissioner for Refugees (UNHCR) displaced from their homes annually from 1990 to 2022. These include people displaced for all reasons, including conflicts, geological disasters (earthquakes, volcanoes), and weather events (floods, storms, etc.) for both those internally displaced (within their own country) and those forced to move to a different country. The numbers are staggering. For the first two decades of this record, they held steady at around 40 million people per year, but since 2013, the numbers have increased dramatically, to a high of 108 million in 2022. Many of these people, especially the internally displaced, have been able to return to their homes during the period shown, but many have not, and will not. Many have died in transit on foot or are amongst the 27,000 victims of overcrowded boats that have capsized in the Mediterranean since 2014 (not to mention other dangerous boat crossings).

While figure 10.1 shows all displaced people, including those uprooted by conflicts, figure 10.2 shows only those displaced to other parts of their home country as a result of natural disasters. Although both geological and weather-related disasters are included, most of these displacements are a result of extreme weather events such as storms, floods, and drought, many of which can be ascribed to the more intense storms and floods and more severe droughts delivered by climate change. Again, the trend shows dramatically increasing numbers over time, notwithstanding large year-to-year variations.

Weather-related displacements take place in all regions of the world, but in 2022—as in most other years—there was a strong concentration of the most consequential events (ones that affected at least hundreds

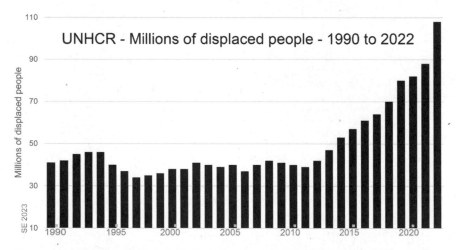

Figure 10.1: *UN High Commissioner for Refugees record of displaced people from 1990 to 2022.*[2]

of thousands of people) in the Global South, including the central part of Africa and in southern and southeastern Asia (figure 10.3). These events include monsoon-related flooding in Nigeria, Pakistan, India, and Bangladesh; devastating tropical storms in the Philippines; and a deadly drought (as many as 43,000 deaths) in Somalia and neighboring countries

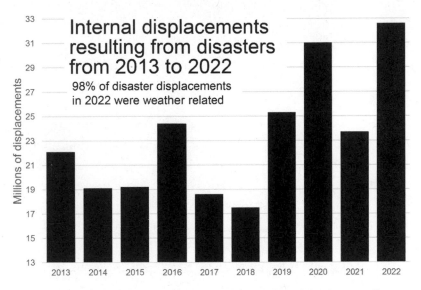

Figure 10.2: *Internal Displacement Monitoring Centre record of people internally displaced from 1990 to 2022.*[3]

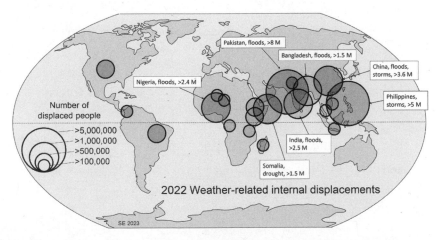

Figure 10.3: *Major weather-related internal displacements in 2022. M = million.*[4]

in the Horn of Africa. They are concentrated in these regions for several reasons: first because these areas are particularly prone to deadly heat, tropical storms, and monsoon variations; second because most of these countries are densely populated; and third because many of the people living there are already vulnerable to displacement for other reasons.

The displacements referred to in figures 10.2 and 10.3 are primarily temporary, resulting from weather-related events, most of which have been exacerbated by climate change. These are not the types of events anticipated in chapter 9 that will make entire regions permanently uninhabitable, but that's only because climate change has yet to advance to that stage. The flood victims of Pakistan, India, and Bangladesh will likely be able to rebuild before the next big flood, although it won't be easy, and the same applies to the storm victims in the Philippines, at least until the next big storm arrives. But this may not the case for the victims of drought in the Horn of Africa region. As summarized in box 10.1, this area has been repeatedly hit by long-lasting rainfall deficits, and while average rainfall amounts may not be decreasing over the long term, higher temperatures are contributing massively to the aridity, making it increasingly difficult to live off the land.

Box 10.1 Horn of Africa Drought 2020 to 2023

In the middle of 2023, countries in the Horn of Africa region (especially Somalia, Ethiopia, and Kenya) are still facing extreme drought conditions after three years of below-normal precipitation. Many people in the region have been living on the edge for decades because of chronic food and water shortages, and limited access to health care, education, and social welfare, and so they are particularly vulnerable to changing conditions. The 2020 to 2023 drought follows severe droughts in 2010-2011 and 2016-2017, and that rapid repetition of multi-year disasters has added significantly to the strain.

In 2022, the drought forced over 2.3 million people in Somalia, Ethiopia, and Kenya to leave their homes and make their way to refugee

Near a refugee camp in Dadaab, Kenya, a young girl stands amid the freshly made graves of 70 children, many of whom died of malnutrition. (Photo by Andy Hall.[5])

centres. An estimated 43,000 died in Somalia alone.[6] A total of 34 million people in the region have been negatively affected, including 5.1 million children that are acutely malnourished.[7]

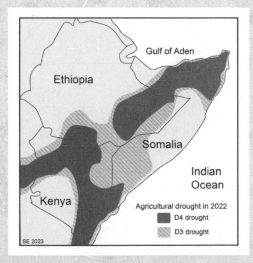

Agricultural drought levels in the Horn of Africa region in 2022.[8]

Drought in the Horn of Africa region is related to oscillating global weather systems, including the ENSO (El Niño-La Niña) variations in the Pacific and the Indian Ocean Dipole, which is characterized by east-west shifts in surface water temperature in the Indian Ocean. Since these are cyclical patterns, there might be grounds for hope that things should get better soon. That may happen , although it doesn't look like it will be this year (2023), so for many more thousands, it will not be soon enough, and it will only last until the next deadly drought.

But the authors of a 2023 report on the region's climate and its potential for food production paint a bleaker picture,[9] one that should give all of us pause. While rains come and go, temperatures continue to rise. Higher temperatures lead to stronger evaporation and so to greater aridity. The drought experienced in 2022 was an extreme (level 4 out of 4) agricultural drought over a wide area in all three countries (see the figure), leading to massive crop failure and the death of livestock by the tens of thousands. And yet, under a cooler climate—such as that which existed 50 years ago, or even 30 years ago—it would not have been an agricultural drought at all. And, of course, as our climate continues to warm, the severity of droughts in this region will only get worse, even with normal precipitation. Within the next decade or two, significant parts of the Horn of Africa region will be uninhabitable because of heat-caused aridity. In other words, they are likely to become part of the stippled area of figure 9.5 long before 2070.

What Is Coming?

Several climate change factors are going to make the issue of human displacement many times more severe in the coming decades than it is now. The first is increasing temperatures, which will have a direct effect on health and life, particularly anywhere that mean annual temperature exceeds 29°C or the wet-bulb temperature ever exceeds 31°C, even for just a few days in a row. Higher temperatures will also exacerbate drought levels and the risk of wildfires. The second is the increasing intensity and frequency of tropical storms and other intense precipitation events on land, and the third is sea-level rise.

As described by Lenton and Xu and co-authors,[10] anticipated climate change to 2070 will cause warming that will render uninhabitable the places where about 3 billion people currently live. The people that survive those conditions will have to go somewhere else. To begin with, the excessive heat will just apply to parts of the affected countries, and many of the victims will become internal refugees (just like those in Somalia, Kenya, and Ethiopia right now). That doesn't mean it will be easy for those victims, or for the people already living in the places where they will be accommodated, but at least it won't involve difficult international borders, or long dangerous journeys overland, or crowded sea crossings in inadequate vessels.

Flooding related to weather events (monsoons and tropical storms) was the most common reason for the internal displacements during 2022 (see figure 10.3). Unlike the victims of drought in Africa, most of those nearly 30 million people may have now returned home and will be in the process of rebuilding their dwellings and farms.

But how long can this rebuilding go on? Over 80 named typhoons (major tropical storms) made landfall in the Philippines from 2000 to 2022, averaging 3 or 4 per year. In 2022 alone, 5 million Filipinos were displaced by tropical storms, bringing the 5-year total (2018 to 2022) to 24 million. This problem isn't going away, and it will only get more extreme as the oceans warm. Within a few decades, it will become impossible to continue living in parts of the country that are repeatedly vulnerable to storm damage. People can rebuild their homes and farms once, twice, maybe three times, but eventually they will make the agonizing decision that they have to find a safe place to live elsewhere. And, of course, the same will apply to many other regions of Asia, much of the Caribbean, and parts of the US Gulf Coast and the Atlantic coast. In some countries, the decisions to abandon a home in a potential flood zone are already being made, not by the homeowners but by the insurance companies that refuse to accept the risk or that charge unaffordable premiums.[11] As the oceans continue to warm, the areas affected by damaging tropical storm winds and flood-inducing rainfall will expand both north and south.

Except on some small Pacific islands, flooding related to sea-level rise has yet to reach the point of large-scale displacements. Of course, this

will change dramatically in the coming decades as sea-level rise acceler-
ates, displacing hundreds of millions from low-relief near-shore regions.

As entire countries become too hot and too dry for survival, or too
frequently flooded by megastorms, or completely submerged by rising
seas, the situation will change dramatically because the victims—perhaps
entire countries' worth of victims—will have to cross borders, deserts,
mountain ranges, and oceans to get to somewhere that they can live.
It is our responsibility to make sure that they are able to do that safely.

There are effectively no legal avenues for climate refugees seeking to
leave their country and enter another one. This issue was raised in the
2015 COP21 Paris Agreement:

> Acknowledging that climate change is a common concern
> of humankind, parties should, when taking action to address
> climate change, respect, promote and consider their respective
> obligations on human rights, the right to health, the rights of
> indigenous peoples, local communities, migrants, children,
> persons with disabilities and people in vulnerable situations
> and the right to development, as well as gender equality,
> empowerment of women and intergenerational equity.[12]

But since then (eight years ago now), there has been no agreement
to provide options for climate migrants. We need an international treaty
that recognizes that increasing numbers of people will no longer be able
to live in their country of origin and will need to move to a place that has
been less impacted by climate change. In the absence of such an agreement, tens of thousands, then hundreds of thousands, and then millions
of climate refugees will either die in place or find themselves beholden
to professional people smugglers, and then to the risk of deportation.

Gaia Vince describes the coming climate refugee crisis in her book
Nomad Century.[13] She emphasizes the "moral abhorrence" of policies
that will close borders to climate refugees or force them into unsafe
boats or deadly desert crossings, and goes on to argue that, because of
the large numbers, such policies would create situations where there is
"no peace for any of us." Vince also argues that, in fact, the prosperous

northern countries, the ones that will be least affected by a warming climate, can actually benefit from large immigrant influxes, as has been the case for well over a century in North America and is being experienced in Europe today. There will be no shortage of work in Europe and North America in the latter half of this century because areas that are intensively farmed now will be too hot and too dry to farm, and eventually to live in, and new farms, new cities, and new infrastructure will have to be built further north.

If we continue to ignore the climate change threat and are not successful in keeping temperatures to less than 2°C above pre-industrial levels, it is quite possible that we will begin transitioning to a PETM-like runaway climate sometime early in the twenty-second century. The rapid feedback related to melting of sea ice and glacial ice will take us over a tipping point that will lead to the release of stored methane in permafrost. That will drive temperatures higher by several degrees, and that will change everything. The areas of Earth that are too hot for humans to inhabit will expand north to encompass much of Asia, all southern Europe, most of the US and even southern parts of Canada, and south into all of Australia and most of the rest of Africa and South America. People living in these places—which is most of us—will be forced to move because it will simply be too hot. Many of the places that are still cool enough to survive in will become too stormy to be safe, and most coastal regions will be flooded. We will be forced to survive in regions that don't have rich soil and where there is very little sunlight for half of each year. It is likely that the Earth's population will decline dramatically because it will be difficult to produce enough food to feed billions of people.

Summary

Even under the most optimistic future emissions scenario, much of the narrative for the rest of this century will be about the displacement of millions of people from places that have become uninhabitable because they are too hot, too dry, too stormy, or flooded with salt water. There are various ways that we can respond to the inevitable reality of millions of climate refugees. One involves closing borders, putting up fences, and sentencing them to die from a crisis that is almost entirely our fault. The much more humane approach would be to open borders and welcome them here in a systematic way, and then encourage them to help us build a new society that is fair for all and respectful of the Earth's limits.

If we can make a sufficiently sharp reduction in our emissions, one that restricts temperature rise to less than 2°C above pre-industrial levels, it may be possible to avoid the strong feedbacks that could tip us over the edge into a PETM-like world. If we cannot, then there will be little that's good for humans in the twenty-second century and those that follow. The parts of the Earth where humans can survive will continue to shrink, our numbers will plummet, and the focus will be on trying to get by in difficult and unfamiliar situations.

The following table provides a summary of how our world will be different if we act quickly to slow climate change and keep temperature change under 2°C versus the possibility that various climate feedbacks described above push us over the edge into a PETM-like runaway climate scenario.

Table 10.1 Summary of the expected impacts of known climate change versus those of PETM-like runaway climate change

Indicator	What is expected with known anthropogenic climate change	What is probable with a runaway PETM-like climate
Air temperature	An increase of 1.5° to 2°C will result in deadly heat waves and serious vegetations stress. It is possible that the temperature could stabilize by the end of this century and might then start to drop.	An increase of 6° to 8°C is possible, making large parts of the Earth uninhabitable by humans. The temperature may not start to decrease for tens of thousands of years.
Terrestrial ecosystems	Floral and faunal communities are already being affected by climate change. That will get worse but may not result in complete ecosystem changes.	There will be wholesale changes to ecosystems making existing landscapes unrecognizable. Farming will be severely restricted by heat and water availability.

Table 10.1 Summary of the expected impacts of known climate change versus those of PETM-like runaway climate change (cont.)

Indicator	What is expected with known anthropogenic climate change	What is probable with a runaway PETM-like climate
Marine ecosystems	Warming oceans will become stagnant and locally anoxic. Coral reefs will disappear from many places but will eventually colonize new areas if temperatures stabilize. Ocean productivity will fall.	Ocean anoxia will be widespread, and deep-ocean ecosystems will suffer significantly. Reefs may be restricted to a few small refugia, and ocean productivity will plummet.
Tropical storms	Tropical storms will continue to increase in number, duration, and intensity, leading to flooding and other destruction in coastal areas and well inland.	Tropical storms will become super-storms, destroying human infrastructure and making many regions uninhabitable.
Sea ice and glacial ice	Almost all sea ice will disappear by the end of this century, along with almost all alpine glaciers. Continental ice sheets (Greenland and Antarctica) will likely survive.	All sea ice will disappear, along with all alpine glaciers. Continental ice sheets will melt completely over the next few millennia.
Sea level rise	Sea level will continue to rise for centuries, but could top out at about 5 meters.	Sea-level rise will continue for millennia and will eventually exceed 65 meters.
Refugees	There will be at least tens of millions of refugees by the end of this century.	Within as little as a century, virtually every human left on Earth could be a climate refugee.

Chapter 11

What Do We Need to Do?

If we gave up eating beef we would have roughly 20 to 30 times more land for food than we have now.

—James Lovelock[1]

THE CLEAR MESSAGE from the preceding chapters is that if we hope to avoid the dangerous climate change that is predicted by climate scientists, or the much more devastating scenario that would come crashing down on us during a PETM-like runaway climate catastrophe, we need to do something, and we need to do it now. Not soon. Now! But what can we do? Is it more important for us all to make individual changes, or for governments to tighten the rules on emissions, or for industry to make big changes? The answer, of course, is all three. Governments control the levers that will bring about major changes, but they are excruciatingly slow to pull them, and the results take a long time to come into effect. Industry, which, with a few exceptions, is motivated almost entirely by profit, will not change until forced to, either by rules or by the market. On the other hand, we control the levers for smaller-scale personal changes, and we can make quick decisions to modify our lives in ways that will have an immediate effect. What's more, we are the market, and we elect the government (in some places at least).

The main sources of our GHG emissions are summarized on figure 11.1. Fossil fuel production and use accounts for 79% of anthropogenic GHG warming on Earth. Obviously, we need to quickly end our use of fossil fuels. Doing that has may co-benefits, as summarized below, and it will take a bonus bite out of our climate impact because it will also curtail most of the fugitive emissions from fossil fuel production.

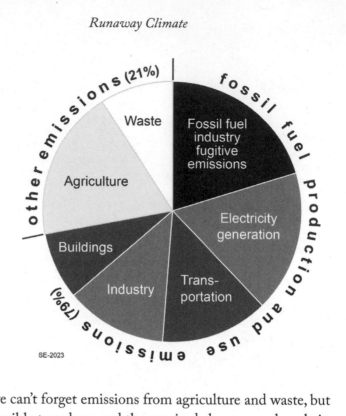

Figure 11.1: *Proportions of anthropogenic GHG emissions derived from our use of fossil fuels and from other sources.[2]*

Of course, we can't forget emissions from agriculture and waste, but those too are possible to reduce, and the required changes are largely in our hands. For more about our GHG emissions please refer to section 4 of the appendix.

Personal Actions

There are some aspects of our personal emissions that we cannot change easily or quickly, such as energy use in our homes, and especially where that energy comes from, but many others are under our control, and can be changed very quickly, at least by individuals, if not by entire communities.

But before we go there, here's something to think about. What if it turns out that climate scientists have got it all wrong, and that people who write books like this one, and speak to community organizations, and organize marches are full of bunk? What if you actually listen to us and make significant changes to your life, like driving way less and walking or biking more, or eating less meat and more vegetables, or renovating your house to make it more energy efficient, and then it turns out you didn't need to do any of those stupid things, and that the so-called climate problem will fix itself? What a bummer! You would end up being fitter and healthier, the air you breathe and the water you drink would be cleaner, some farmland now used to feed or raise beef cows might get returned to forests or even parks, and your home would be more comfortable and cheaper to live in—all for no damn good reason! The message, of course, is that many of the personal changes that we must make to reduce our climate impacts can be win-win-win changes that will also benefit us and our immediate environment. In other words, they are changes that we should embrace regardless of our concern for the climate. We'll focus on some of those.

How about walking or biking to work, or to shop, or to college? That's so much better for your physical and mental health than sitting in a car and swearing at the traffic! If you're not ready to pedal up those hills, an electric bike can help. You'll save a lot of money on fuel and vehicle maintenance. You might qualify for a reduced insurance rate if you're not driving your car to work every day, or you might even be able to divorce your car. You'll get to feel the seasons go by (the good and the bad!) and experience how your community changes from month to month. Some of the benefits of switching modes of transportation (and other solutions) are summarized in table 11.1.

Table 11.1 Some of the co-benefits of personal changes to reduce GHG emissions

Change	Climate benefits	Other personal and community benefits
walking or biking more	significantly reduced GHG emissions	improved physical and mental health, cleaner air, less traffic congestion, less parking needed, substantially reduced transportation cost
using transit more	reduced GHG emissions	improved physical and mental health; extra time to work, read, or scroll; cleaner air; less traffic congestion; less parking needed; reduced cost

Table 11.1 Some of the co-benefits of personal changes to reduce GHG emissions (cont.)

Change	Climate benefits	Other personal and community benefits
switch to an EV	reduced GHG emissions	cleaner air, quieter streets, lower overall cost to drive
home efficiency upgrade	reduced GHG emissions in most places	lower heating and cooling costs, healthier and more comfortable living spaces
install a heat pump	reduced GHG emissions in most places	lower heating and cooling costs, potential for cooling during heat waves
install solar modules	reduced GHG emissions in most places	lower electricity costs, reduced strain on the electrical grid, electricity available during power outages if you also install battery backup
eat less beef	significantly reduced methane emissions, increased CO_2 sequestration	improved health outcomes, lowered body mass index, lower food cost, multiple benefits from rewilding farm land
avoid organic waste	reduced methane emissions	lower food costs, improved soil fertility

OK, not ready to walk or bike. How about catching the bus? First, you'll have to walk a distance to the bus stop, and maybe at the other end as well. That little bit of walking, in fact, is one of the greatest health benefits of using transit. The whole trip might take a little longer, but you'll get most of that time back because you can do things on the bus that you can't do while driving. It will almost certainly cost you less per trip. Yes, transit schedules and routes are not what they could be—especially in North America—but if more people started using buses and trains, transit would get better.

Transit doesn't work for you? How about an electric car? Yes, they are expensive, but that cost will be repaid in reduced fuel and maintenance costs. EV costs are coming down, and within a decade or so, you won't be able to buy a new fossil-fueled car anyway. In addition to reducing GHGs, faster adoption of EVs will also make our streets quieter and the air easier to breathe. An illustration of the dramatic public health benefits of reducing local pollution that we could gain by curtailing our use of fossil fuel powered vehicles—through more cycling and walking, greater use of public transit, and the switch to electric vehicles—is provided in box 11.1.

What about where you live? Is your home energy efficient? Is it bigger than what you really need? Could you benefit from moving closer to work or school or shopping? Does your roof have solar potential?

Box 11.1 The Public Health Benefits of COVID-19 and of Taking on Climate Change

There is no doubt that the COVID-19 pandemic has had, and still has (in mid-2023), enormous health, social, and economic costs, but there have been some surprising beneficial outcomes. Who doesn't love virtual family gatherings and day-long Zoom meetings? OK, maybe you don't, but you'll be glad to hear that there was a significant but temporary reduction in CO_2 emissions during the height of the COVID outbreak in 2020, which resulted in small overall reductions in CO_2 emissions for the year—a drop of 6.4% from the previous year[3] (see figure 7.4). The decline in the use of fossil-fueled vehicles also led to reduced emissions of other types of pollutants, especially nitrogen oxide and particulate matter.[4] Some benefits of that were clearer views and brighter night skies in many places that are normally quite badly polluted, but there were also significant health benefits.

In China the use of private vehicles was strongly curtailed by the government over a period of several months in 2020. That led to a dramatic reduction in vehicle-related emissions including NO_2 and particulate matter. Yale University's Kai Chen and co-authors estimated that the number of deaths *avoided* because the air was cleaner due to the driving restrictions during that period was just over 12,000, while the number of deaths attributed to COVID during the same period was around 4,300.[5]

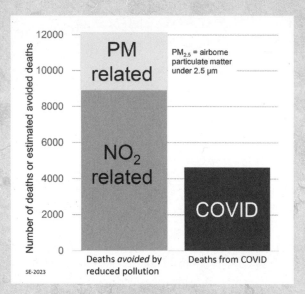

Based on data in
Chen, K., et al.[6]

Surely, we don't need a pandemic to force us into changing our transportation modes so that we can all benefit from positive health outcomes! We just need to reach the inevitable conclusion that driving fossil-fueled vehicles is very bad for our collective health and for the planet's health. And that's not just because of the pollution they cause.

How does taking on climate change impact our health overall? The authors of a recent study published in *The Lancet* added up the expected health impacts of the UK's plan to be carbon neutral by 2050.[7] They point out that carbon-reduction measures benefit the health of a population because the air will be cleaner, people will be more physically active, homes will be better insulated and so more comfortable, and people will eat less meat and dairy and more fruit and vegetables. They estimate that the most aggressive planned pathway to net-zero by 2050 will result in over 13 million life-years gained by the population of England and Wales by 2100.

If your home is older, you could save a lot of energy by upgrading with better windows, reducing leakage with weatherstripping and caulking, or by adding insulation to the walls and attic. You could install a more efficient heating and/or cooling system, such as a heat pump. Many jurisdictions already have incentives to make such changes. If much of your electricity comes from fossil-fuel sources (which still applies in most states, some provinces, and many countries), then putting solar modules on your roof can make a big GHG difference and save you money. Solar electricity is cost-effective in most parts of North America and Europe, even in places where it seems to be raining much of the time.[8] The benefits and co-benefits of some of these options are summarized in table 11.1.

Maybe your diet is part of the problem. As you can see from figure 11.2, a diet that is rich in red meat (especially beef) adds a lot to your GHG production, more so than any other food choices you make. Reducing your meat intake would immediately reduce your carbon footprint, and your food budget. It would likely make you healthier too. A review study carried out in 2015 shows a strong correlation between beef consumption and negative health outcomes.[9] If most of us were to significantly reduce our beef consumption, our GHG emissions would go way down, and it might also be possible to restore the forest on some pasture and grain farmland. That's because we could grow equivalent amounts of protein embodied in other foods on a small fraction of the land that it takes to produce beef. Over 165 square meters (m^2) of land is needed to produce 100 grams of protein as beef, versus 40 m^2 for cheese, 27 m^2 for milk, 8 m^2 for nuts, and 2 m^2 for tofu.[10] See section 5 of the appendix for an explanation.

The "waste" slice in figure 11.1 represents emissions from both solid waste (landfill waste) and sewage. There isn't much that individuals can do about their sewage waste (except maybe to eat less!), but we can make a difference with our landfill waste. Remember that the key problem is the emission of CO_2 and methane that results from the breakdown of organic matter within landfills. Our main goal needs to be reducing the amount of organic matter that gets placed in a landfill. Some municipalities have compost collection systems in place, so residents of those places must make sure to put their compost into the green bin provided.

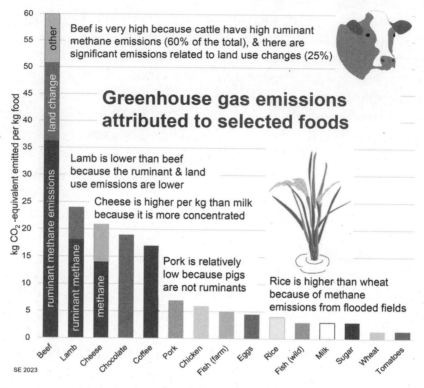

Figure 11.2: *Greenhouse gas emissions attributed to various food types.*[11]

In places where such programs don't exist, residents should do everything they can to limit the organic component of their waste. That might start with preparing less food so there is less wasted food on plates and making sure that leftover food gets used. For those that can, it should also mean backyard composting so that the carbon in organic matter can be sequestered in the soil. Those that cannot compost at home might be able to find community composting options in their neighborhood. Food waste is a problem because no matter what we do with it, there will be some GHG emissions. If it is sent to a regular landfill, a significant proportion will be as methane, and that needs to be avoided. At most composting plants, nearly all emissions are as carbon dioxide, not methane.

To sum up, most of the personal changes that we need to make to reduce our climate impacts have positive side effects that will make us physically and mentally healthier and also more comfortable in our homes. They will also have significant other environmental benefits. These may seem like small actions that will have little impact on a big problem, but this is exactly how big things get done. If you set the example, individual actions can snowball through your community.

One other issue, and perhaps it is the most important individual contribution that we can make, is to vote every chance we get. Voter turnout is depressingly low in many places around the world for several reasons, the mostly commonly cited ones being "not interested in politics" and "too busy." Both stem from a lack of confidence in, or even outright disdain for, political systems and politicians. While that is perfectly understandable, if we want to see changes in how things are done— changes that will lead to a viable climate future—we need to get over it. According to the business magazine *Forbes*, the main reason Donald Trump won the 2016 US election was because Democrat voters stayed home in droves on election day.[12] So, if you care about the future, you need to hold your nose and vote, you need to tell your friends and family that you voted, and you need to let the politicians know that you are voting for the climate (and of course for whatever else you think is important).

If you are curious to know about your own GHG emissions, you could use one of the on-line emission calculators or the quick-and-dirty calculation method provided in section 6 of the appendix.

The Role of Governments

As important as they are, and as beneficial they might be in many unintended ways, personal behavior changes will not be enough to keep us out of climate trouble. We need major government-led incentives to encourage people to change and strong regulations to force institutions and corporations to change.

Many governments don't take any meaningful action on climate change because they assume that most of their voters don't see it as a priority. That may be the case in some regions, but it is gradually changing, and climate change is increasingly one of the top issues on the minds of most voters.

If and when we get governments that are genuinely open to actually taking action on climate change (rather than just talking about it), there are many issues that we need them to start working on right away.

Fossil Fuel Subsidies

The first priority is getting governments to stop encouraging and supporting the fossil fuel industry. Existing subsidies are huge (figure 11.3) and have various forms, including direct cash grants, reduced royalties, relaxation of environmental regulations, tax credits, loans or loan guarantees, and government funding of infrastructure such as pipelines. Governments say they subsidize fossil fuels to protect jobs in that industry and to protect consumers from high fuel prices, but it's more about repaying their friends in the industry who have financially supported their election. Fossil fuel subsidies just make fuel seem artificially cheap so that people use more, and they penalize those who use less.

The fossil fuel industry no longer deserves government support of any kind. It is a sunset industry made up of corporations that have been lying to us for decades and are seriously harming the planet. Public money would be much better spent on subsidizing non-fossil energy projects like wind and solar, electrical grid enhancements to promote energy

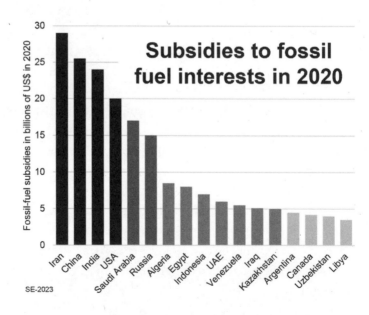

Figure 11.3:

National-level

subsidies to

fossil fuel

interests

in 2020.

Sub-national

governments

also provide

subsidies.[13]

sharing across regions, and energy storage systems that support wind and solar generation and also enable reductions in peak electricity production.

Zero Carbon Commitments

The 2015 Paris Agreement, to which almost every country in the world is a party, includes an aspiration to achieve global net-zero carbon emissions by 2050. Over the several years since then, many countries have made pledges or created laws or policies to meet that target. The status of those commitments for the 24 major carbon-emitting countries is illustrated in figure 11.4. A few countries (Germany, Sweden, Finland, Austria, and Iceland) have net-zero targets earlier than 2050, while some very significant emitters (Brazil, China, India, and Russia) have later targets. The commitments do not add up to net-zero emissions by 2050 because of the large overall contributions of China and India.

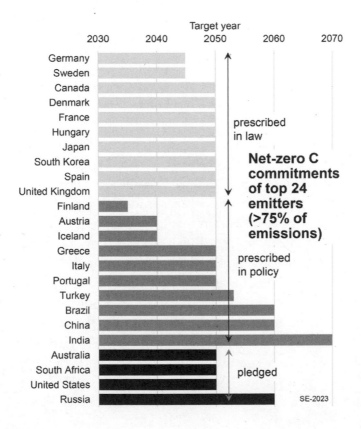

Figure 11.4: The net-zero commitment status of the top 24 emitting countries, as of November 2021.[14]

It is important to remember that these are "net-zero carbon goals," not "zero carbon goals." Most countries have included carbon sequestration projects (such as reforestation, often in some remote and virtually invisible places) as part of their arithmetic. As revealed in a January 2023 study by the *Guardian* and *Die Zeit*,[15] many of these carbon-offset projects are not achieving carbon sequestration close to the rates alleged. The investigators concluded that 94% of the carbon offset credits claimed at the 29 sites they studied did not meet prescribed standards and should not have been approved. Because a lot of money can be made by selling carbon offsets, there is an incentive to exaggerate—or simply lie about—their effectiveness. The lesson is that we need to be extremely careful about accepting offset claims and should insist that governments and corporations to make real zero carbon commitments, without relying on offsets.

Ending Fossil Fuel Development

Fossil fuel extraction and distribution projects typically have lifespans of several decades, and since most countries have pledged or committed to get off fossil fuels by 2050, it makes absolutely no sense to start new fossil fuel projects now. Some countries, including France, Spain, Denmark, Ireland, Greenland, Belize, and Colombia, have already banned further exploration and production, and while this doesn't have big implications for France and Spain, which have very limited fossil fuel potential, it is a big deal for Belize and Colombia, which have significant potential.

The United States and Canada are both major fossil fuel producers, and neither country has made any commitments to limit new development projects. In March 2023, President Biden torpedoed any good that his other policies might have accomplished by approving the Willow oil drilling project in Alaska. Canada is currently providing a huge incentive to the fossil fuel sector by continuing construction of a second Trans-Mountain pipeline from northern Alberta to the Pacific coast near Vancouver. The pipeline will triple the rate of the overseas export of dirty tar sands oil. Why is the government of Canada pouring money into a project that supports production of some of the dirtiest oil on the planet, isn't wanted now, and will not be needed at all in 25 years?

Putting a Price on Carbon Emissions

Greenhouse gas emissions have direct climate effects that represent huge costs to the environment, to people, and to infrastructure. If we accept that polluters should pay for the costs of their pollution, then it is obvious that we need to set a fee for doing so, first to recover some of that cost and second to discourage pollution, especially climate change pollution (greenhouse gases). There are two main ways of pricing carbon emissions: one is through a carbon tax, a direct tax on CO_2 emissions, and the other is through an emissions trading system (ETS, also known as cap-and-trade) that sets an upper limit on the emissions within a jurisdiction and/or sector and allows entities that emit at rates under the limit to accrue credits and then sell them to entities that emit over the limit.[16]

A carbon tax is typically charged on fuel at the pump, and normally extends to everyone—including individuals and corporations. Carbon tax can also be applied to less direct emission sources, such as beef and dairy herds. The current prices vary widely, from well under $1/tonne of CO_2 (in numerous countries) to nearly $130/tonne in some northern European countries; a global average is close to $2/tonne. In many places, carbon taxes started out quite low and have been gradually increased to put pressure on consumers to continue reducing their emissions. Carbon taxes are effective because they drive shifts towards low-emission behaviors and technologies. In some jurisdictions, a portion of carbon tax revenue is returned to citizens in the form of rebates, and for the majority, these rebates actually exceed the amount they pay in tax. That is especially true for low carbon emitters (who pay less tax). Only the higher carbon emitters pay more tax than they receive in rebates.

An ETS, or cap-and-trade system, typically only applies to industry, and the price of carbon credits varies depending on their supply and demand. As entities in a region or sector collectively approach the cap, the price will increase. The price of carbon in an ETS system is therefore subject to changes in the cap level and to how well the sector performs on reducing emissions. If the cap is reduced, the availability of credits will drop, and the price will go up, but if entities within the sector make broad and effective emission cuts, the price will drop. ETS prices currently vary from under $1/tonne of CO_2 to approximately $100/tonne.

As of 2021, carbon pricing systems (C-tax and/or ETS) were in place in over 40 countries, and they covered approximately 25% of global emissions. As Ingo Venzke noted: "The destruction of the planet remains hugely profitable, and legally protected."[17] That needs to change, and so the price of carbon needs to go up, way up! In order to cover even part of the social, economic, and environmental costs of carbon emissions, the average price should be over $100/tonne, and in order to be effective in keeping us to the Paris Accord target of 1.5°C, that will have to increase to over $1,000/tonne by 2050. For some context, $1,000/tonne is equivalent to about $2.30/L (or $8.75/gallon) of fuel at the pump.

There is widespread agreement amongst economists, scientists, and policy makers[18] that carbon pricing is an effective and fair way to reduce our climate impacts and keep us out of climate jeopardy, but taxes of any kind are a hard sell, especially in the US, so politicians are hesitant to get on board. Voters need to tell their politicians and political candidates that they support a price on carbon, and then they need to start reducing their own emissions so they don't get hammered.

An alternative way to drive emission reductions would be to establish a personal carbon-emission rationing system. Such a system would limit personal annual carbon emissions, but unlike the cap-and-trade system used to limit industrial emissions, personal emissions rations would not be tradable.[19] This would ensure that everyone had equal access to energy because the wealthy could not simply buy as many credits as they want from the less wealthy. Some might argue that it makes no sense to ration something that isn't scarce. Fossil fuels are not scarce, but the capacity of the atmosphere to absorb more carbon without significant climate risks is very scarce, and so it would be our right to dump carbon into the air that is being rationed, not our right to buy fuel or take holidays on other continents.

Encouraging Personal Changes

Our governments can play an important role in helping more of us reduce our personal emissions, and it doesn't have to cost the Earth. Much of it can be achieved by redirecting existing government resources into areas that will help us drive less, drive smarter, be more active, and eat better. Here are some of the things that governments should do to help us.

- Spend less on ever wider and faster highways, bridges, and tunnels. The response to more road capacity is always more traffic because people start making trips that they would not have made before (and so presumably didn't need to make), they make longer trips, and they shift to driving from other modes.[20] Yes, it's true that drivers stuck in traffic are always clamouring for bigger roads. Smart governments ignore that clamour and instead provide better alternative transportation infrastructure.
- Spend more to improve transit. It could be as simple, quick, and inexpensive as giving buses priority, or providing more buses, but it can also involve larger investments, such as building grade-separated rail systems. We need to get people out of cars, but it won't happen if transit isn't a viable option.
- Build better infrastructure for walking and biking. That can be expensive too, but nothing like building freeways, and in some cases, existing roadways can be repurposed for use by pedestrians and cyclists. In 2009, the City of Vancouver gave over one of the six traffic lanes on the Burrard Bridge to cyclists. By 2019, another lane of traffic has been dedicated for cycling and pedestrians, the bridge had become the busiest bike route in North America, motor-vehicle crossing times were no slower than before, and downtown businesses—that had originally opposed the move—embraced it because their employees and customers were arriving by bike.[21]
- Provide incentives to accelerate the uptake of electric vehicles. Norway is the uncontested leader in EVs; they made up just shy of 80% of all vehicle sales in 2022, and that's because the country provides multiple incentives to purchasers.
- Provide education about and incentives for home upgrades, including switching to heat pumps, upgrading windows, and improving insulation.
- Encourage a shift away from meat-heavy diets (especially beef). That can start with an end to subsidizing the beef and dairy

industries (reported to be in the order of $38 billion per year in the US[22]), and it can continue through carbon taxes. Prices will then reflect the actual cost of meat production to the planet, and meat consumption will drop. Far from being a cost to governments, this type of change will save on subsidy payments, increase income from carbon taxes, and dramatically reduce health-care costs.

Summary

Individuals, governments, and corporations all have a role to play in keeping us out of climate trouble and getting the world to actual net-zero emissions by 2050. If we can achieve that, we should prevent some of the worst outcomes of steady-state climate change, and we might be able to avoid sliding over the edge into a PETM-like runaway climate crisis.

The most important personal changes are driving less—much less—flying less, walking and cycling more, using transit if walking and biking are not practical, and switching to an EV when we have to drive. We also need to make our homes more efficient, and that could include living in less space, upgrading windows and insulation, installing a heat pump, and, where practical, putting up some solar panels. Many of us need to make significant changes to our diets by cutting way back on beef and dairy and eating more vegetables. The good news is that all these changes will save us money and make us healthier and more comfortable in our homes. The other key thing we need to do is vote, and we need to let candidates and political parties know that we are voting for the climate.

Governments have the most important role to play when it comes to climate change because they can persuade both individuals and corporations to make changes.

- First and foremost, they need to immediately end financial and regulatory support for the fossil fuel industry, support that allows corporations to provide big dividends for their investors and make fossil fuels artificially cheap, thus driving up consumption.
- Second, they need to make strong and achievable commitments to reach net-zero carbon emissions by 2050, to ensure that most of that is based on actual cuts to emissions and not on questionable

offsets, and then to systematically strengthen and shorten those commitments over the next two decades.

- Third, they need to immediately end all new fossil fuel developments, and then work with the fossil fuel corporations to transition to socially and environmentally beneficial endeavors before 2050.
- Fourth, they need to put a price on carbon, either with a carbon tax or an emission trading system, and then gradually adjust those measures so that the cost of emitting carbon matches the actual cost to the environment.
- Fifth, they need to create programs and build infrastructure that will help and encourage individuals to make the changes outlined in the paragraph above.

List of Abbreviations

Term	Page	Meaning
^{14}C, ^{13}C, ^{12}C	38	Carbon-14, -13 and -12—three isotopes of the element carbon
^{18}O, ^{16}O	35	Oxygen-18 and oxygen-16—two isotopes of the element oxygen
$CaCO_3$	32	Calcium carbonate, or calcite (although other forms of $CaCO_3$ exist)
CCD	33	Carbonate compensation depth—the depth in the ocean below which the mineral calcite is soluble (approx. 4,500 m in most parts of the ocean)
CH_4	21	Methane (a GHG)
CIE	41	Carbon isotope excursion—a change over time in the average ratio of ^{13}C to ^{12}C
CO_2	21	Carbon dioxide (a GHG)
COVID	152	The SARS coronavirus infection that started in late 2019 and became a global pandemic
ENSO	140	El Niño southern oscillation—the Pacific Ocean current and temperature changes associated with the El Niño and La Niña phenomena
ETS	159	Emissions trading system—an economic protocol for controlling industrial carbon emissions to the atmosphere
GCM	42	General Circulation Model—a digital model that can be used to understand past or future changes in the Earth's climate
GHG	21	Green house gas—a gas that can trap heat in the atmosphere

Term	Page	Meaning
HFC	169	Hydro-fluoro-carbon—a gas comprised of hydrogen, fluorine and carbon that is used in cooling systems and is also a GHG
IPCC	44	Intergovernmental Panel on Climate Change
K-Pg extinction	10	Cretaceous-Paleogene extinction event—the dramatic environmental changes at 66.0 Ma that led to the demise of the dinosaurs
Ma	6	Mega annum—Millions of years ago
MAT	124	Mean annual temperature—the average annual temperature of a place on the Earth, or of the Earth as a whole
NADW	113	North Atlantic Deep Water—cold salty water in the northern Atlantic Ocean that sinks to become part of a deep ocean current
NASA	120	US National Aeronautics and Space Administration
ODP	53	Ocean Drilling Project
PETM	ix	Paleocene-Eocene Thermal Maximum—a 180,000 year period, starting 56.0 million years ago, during which the Earth's average temperature increased by around 7° C
POE	68	Pre-onset excursion—a short CIE that preceded the main PETM CIE by several thousand years
SLR	130	Sea level rise—the global rise in sea level that is the result of human-caused climate change
SO_2	26	Sulphur dioxide (not a GHG)

Appendix

1 Weathering and Erosion

EARTH IS AN ACTIVE PLANET. The core is hot enough to generate convection in the mantle, and that slow churning action is the primary driver behind movement of the tectonic plates. The plates move in different directions and at different rates. Processes that take place at plate boundaries lead to earthquakes, volcanoes, and the formation of mountain ranges. If not for that, the Earth's surfaces would be deadly flat. As long as there are mountains, there is erosion. Gravity, precipitation, and weathering slowly turn immense masses of solid rock into sand, clay minerals, and ions in solution. Those are washed into the oceans by rivers, and they accumulate on the ocean floors as sedimentary layers.

The process of chemical weathering is illustrated in figure A.1. The fresh surface of the granite rock (on the left) has bright white feldspar, glassy quartz, and black amphibole. On the weathered surface of the same rock (right), the feldspar and the amphibole are chalky beige and dark grey, respectively, while the quartz remains glassy (although that's difficult to see here).

In the chemical weathering process illustrated in figure A.1, feldspar reacts with carbonic acid (made from H_2O and CO_2 in the atmosphere) and oxygen to form the clay mineral kaolinite, while calcium and carbonate ions are washed away in surface water.

plagioclase feldspar + carbonic acid + oxygen $\longrightarrow$ *kaolinite +*
calcium ions + carbonate ions[1]

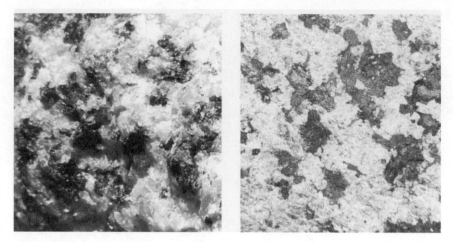

Figure A.1: *Two views of the same piece of granite. Left, a recently broken fresh surface; right, a weathered surface. Both images are about 2 centimeters across.*

The conversion of hard feldspar into soft kaolinite weakens the rock. Eventually it breaks apart into sand-sized grains of quartz plus some feldspar and amphibole, and kaolinite. The strong, hard, and chemically inert quartz is highly resistant to further breakdown—which is mostly why we have sandy beaches—while the feldspar and amphibole, which are weaker minerals, gradually get broken into smaller pieces and then into more clays, releasing other ions into the water, including those of iron, magnesium, sodium, potassium, and silicon. The clay minerals get carried away in surface water.

Importantly, weathering of silicate minerals, such as feldspar, consumes atmospheric CO_2 and delivers the carbon to the oceans, where it can be stored for a very long time. The reduction in CO_2 levels has a cooling effect on Earth's climate.

2 Changes to Floral Communities During the PETM

Figure A.2 shows some of the floral community changes observed in Paleocene and Eocene rocks in the Bighorn Basin. Some plants were extirpated from the region at (or just before) the start of the PETM and did not reappear after. Some were extirpated and did return after, some remained in the area throughout the PETM, and others (at the bottom) were only present during the PETM.

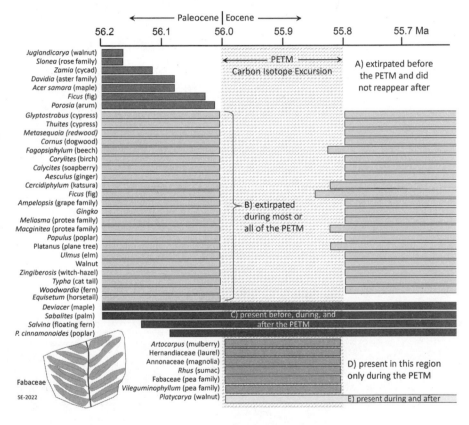

Figure A.2: *Known floral extirpations and immigrations in the Bighorn Basin region during the PETM. (Genus names are in italics. Fossil sites representative of some of these time periods are limited, and it is possible that some of these plants were not present in the basin throughout the time intervals shown.)[2]*

3 Changes to Faunal Communities During the PETM

Figure A.3 shows some of the faunal community changes observed in Paleocene and Eocene rocks in the Bighorn Basin. Some animals became extinct just before the start of the PETM and did not reappear after. Some remained in the area throughout the PETM, and others (at the bottom) appeared in the area near to the start of the PETM, and remained after.

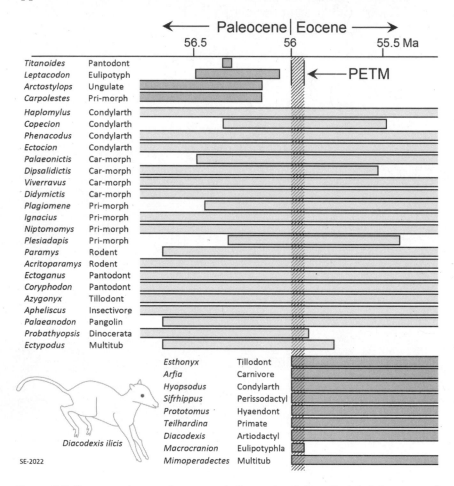

Figure A.3: *Genera and groups for representative mammals from the late Paleocene and early Eocene. Pri-morph = Primatomorpha, Car-morph = Carnivoramorpha, Multitub = Multituberculata , Hyaendont. = Hyaenadontidae; see table 5.1 for further explanation.[3]*

4 Climate Drivers

Figure A.4, which is based on information in the 2021 IPCC report, provides an overview of the ways in which we are affecting the climate through emissions of greenhouse gases and other activities. The main greenhouse gases, which lead to warming, include CO_2 and methane (CH_4) along with nitrous oxide (N_2O) and hydrofluorocarbons (HFCs) and the water in jet contrails. We also need to consider aerosols and soot, which actually contribute to cooling by blocking incoming sunlight.

When considered over a short time period—10 years—the largest contributors to climate warming are leakages and other emissions from the fossil fuel industry, followed closely by emissions from agriculture. In both cases, the key GHG is methane (see box A.1). There is also a small cooling effect from aerosols that are emitted in the process. Methane is also the most important emission from landfills. Heating buildings with oil or gas produces CO_2, while leaks from refrigerating devices (air conditioners, refrigerators, and freezers) contribute HFCs. Again, there is a small cooling effect from aerosol emissions. The other main uses of fossil fuels are electricity production; cars, trucks, and trains; and industry (including construction and manufacturing). All of these emit CO_2, and so contribute to warming, and they also have quite significant cooling effects from smoke, ash, and sulphate aerosols. The shipping industry emits a relatively small amount of CO_2 and has a

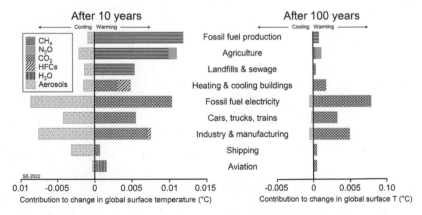

Figure A.4: *Surface air temperature change effect of one year's worth of present-day emissions from various sectors, after 10 years (left) and after 100 years (right).*[4]

sizeable short-term cooling effect (because of particulate pollution from the low quality of fuel used on ocean vessels). The aviation industry has a relatively small warming effect overall, from CO_2 and from injection of H_2O into the stratosphere.

Box A.1 Absorption Windows

Methane (CH_4) is a more effective GHG than CO_2 for the reason illustrated in the diagram.

The Earth (warmed by the sun) emits radiation in the infrared (IR) part of the spectrum (mostly in the range from 5 to 20 μm [micrometers]). Much of that radiation is absorbed by atmospheric water vapor (at wavelengths in the hatched regions), but IR radiation still gets through within two important windows (the white areas) and continues on through the atmosphere.

CH_4 absorbs much more strongly than CO_2 within these windows. In fact, molecule for molecule, CH_4 is about 100 times as effective at absorbing IR radiation as CO_2.

But CH_4 isn't stable in the atmosphere; it gradually gets oxidized to CO_2 and H_2O. After about 10 years, half of the CH_4 molecules will be gone, and after 100 years, virtually all will be gone. On the other hand, a molecule of CO_2 can remain in the atmosphere for centuries.

Based in part on public domain diagram by R. Rohde.[5]

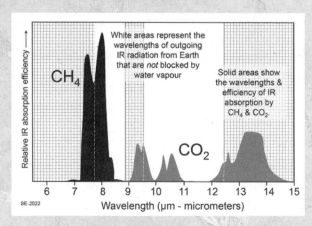

5 Land Use for Beef Production

Figure A.5 illustrates the amount of land that is required to produce protein by growing different types of food. The important point is that not only is beef production highly carbon intensive (see figure 11.2), it also requires a great deal of land. We would be much better served if the land currently used to raise beef (including the separate land where cattle feed is grown) was converted either to other types of food production or to carbon-sequestering forest.

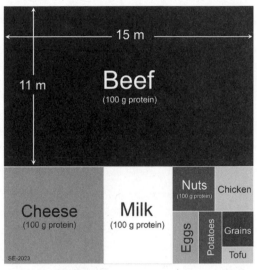

The diagram shows how much land is needed to grow 100 g of protein with different types of food. 100 g of protein from potatoes requires about 3% as much land as 100 g of protein from beef.

Expressed another way, if the 15 by 11 m plot of land that is needed to produce 100 g of beef protein was converted to some other type of production, we could instead produce one of the following amounts of protein:
412 g in cheese, 605 g in milk, 2066 g in nuts, 2324 g in chicken, 2902 g in eggs, 3168 g in potatoes, 3565 g in grain, or 7455 g in tofu.

Careful, though, because some land that is used for beef grazing is not of sufficient quality to produce protein using other crops at the rates listed.

Figure A.5: *The amounts of land needed to produce 100 grams of protein.*[6]

6 Estimating Your Carbon Footprint

Curious about your carbon footprint? You can go on line to find out where you stand,[7] or you can do a quick and dirty calculation right here with the help of the figure below. Before you start, just bear in mind that this is for a one-year period (the last twelve months will do, or the last calendar year, if you prefer), and that it's just for you, not for your entire household.

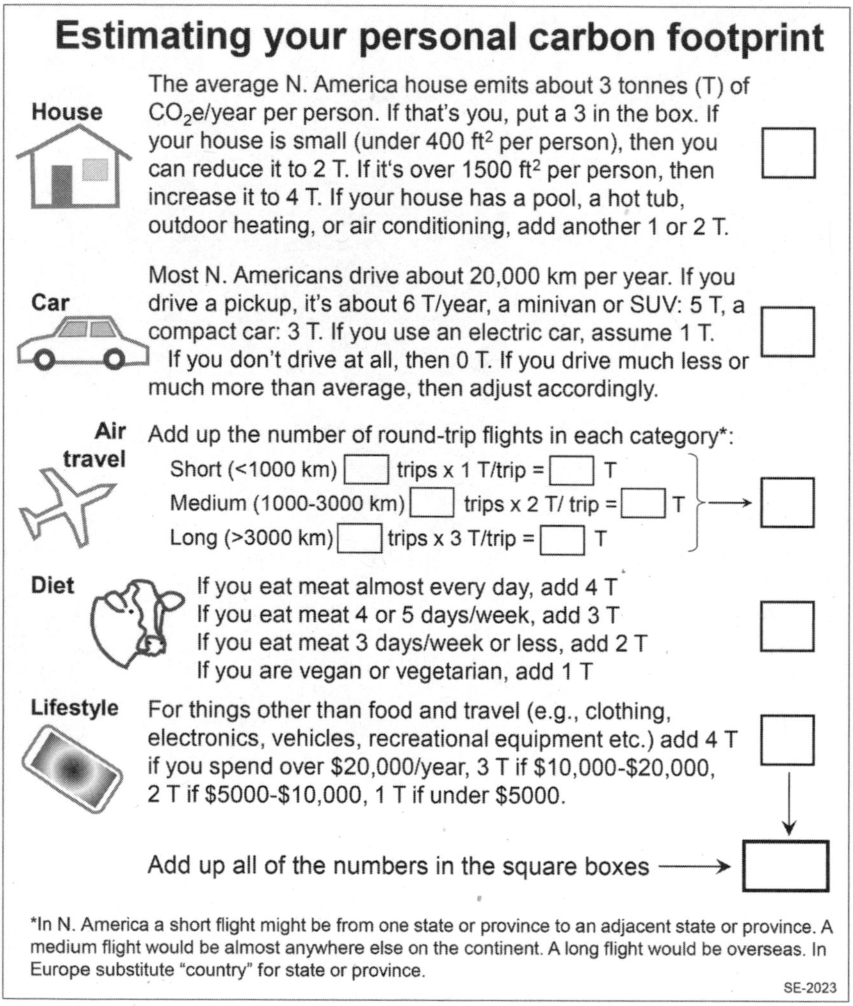

Estimating your personal carbon footprint

House

The average N. America house emits about 3 tonnes (T) of CO_2e/year per person. If that's you, put a 3 in the box. If your house is small (under 400 ft² per person), then you can reduce it to 2 T. If it's over 1500 ft² per person, then increase it to 4 T. If your house has a pool, a hot tub, outdoor heating, or air conditioning, add another 1 or 2 T.

Car

Most N. Americans drive about 20,000 km per year. If you drive a pickup, it's about 6 T/year, a minivan or SUV: 5 T, a compact car: 3 T. If you use an electric car, assume 1 T. If you don't drive at all, then 0 T. If you drive much less or much more than average, then adjust accordingly.

Air travel

Add up the number of round-trip flights in each category*:

Short (<1000 km) [ ] trips x 1 T/trip = [ ] T

Medium (1000-3000 km) [ ] trips x 2 T/ trip = [ ] T ⟶ [ ]

Long (>3000 km) [ ] trips x 3 T/trip = [ ] T

Diet

If you eat meat almost every day, add 4 T
If you eat meat 4 or 5 days/week, add 3 T
If you eat meat 3 days/week or less, add 2 T
If you are vegan or vegetarian, add 1 T

Lifestyle

For things other than food and travel (e.g., clothing, electronics, vehicles, recreational equipment etc.) add 4 T if you spend over $20,000/year, 3 T if $10,000-$20,000, 2 T if $5000-$10,000, 1 T if under $5000.

Add up all of the numbers in the square boxes ⟶ [ ]

*In N. America a short flight might be from one state or province to an adjacent state or province. A medium flight would be almost anywhere else on the continent. A long flight would be overseas. In Europe substitute "country" for state or province.

SE-2023

Figure A.6: *Template for a quick and dirty estimation of individual carbon emissions*

The emissions for an average North American are about 15 T/y (tonnes of CO_2 equivalent per year). That's much higher than the European average of close to 7 T/y, the global average of around 5 T/year, or the African average of 1 T/y. If your emissions are over 5 T/y, then you are part of the problem. If they are over 15 T/y, then you are a major part of the problem! When looking at your numbers, think about some of the changes you could make that would contribute most to decreasing your climate impact.

Endnotes

Chapter 1

1. Marsh, O., 1877, *Introduction and Succession of Vertebrate Life in America*, Tuttle, Morehouse, & Taylor.
2. Much of the following is summarized from Johnson, K. and Clyde, W., 2016, *Ancient Wyoming: A Dozen Lost Worlds Based on the Geology of the Bighorn Basin*, Fulcrum and the Denver Museum of Nature and Science.
3. In this volume, genus names like *Psilophyton* are capitalized and italicized. Species names (e.g., *P. dawsonii*) are italicized but not capitalized. Higher level names of organisms (e.g., families, orders, classes) are capitalized but not italicized. For example, Eulipotyphla is an order within the class Mammalia.
4. Scotese, C., 2021, "An Atlas of Paleogeographic Maps: The Seas Come In and the Seas Go Out," *Annual Reviews of Earth and Planetary Sciences*, 49, 669–718. All of the maps in this volume that portray past configurations of continents and oceans are based on the work of Christopher Scotese, as summarized in "Atlas of Paleographic Maps." See www.scotese.com for more information and more maps.
5. Paleogene is the period (like Cretaceous). Paleocene, Eocene, and Oligocene are the epochs of the Paleogene.
6. The current rate of anthropogenic temperature change is approximately 6,700 times faster than that.
7. Most of the previous paragraph is based on the findings of Wilf, P., et al., 1998, "Portrait of a Late Paleocene (Early Clarkforkian) Terrestrial Ecosystem: Big Multi Quarry and Associated Strata, Washakie Basin, Southwestern Wyoming," *Palaios*, 13, 514–32.

8. Jarvis E., et al., 2014, "Whole-genome Analyses Resolve Early Branches in the Tree of Life of Modern Birds," *Science*, 346, 1320–31.
9. *Wikipedia*, "Sifrhippus."
10. Gingerich, P., 1989, "New Earliest Wasatchian Mammalian Fauna from the Eocene of Northwestern Wyoming: Composition and Diversity in a Rarely Sampled High-floodplain Assemblage," *University of Michigan Papers on Paleontology*, 28, 1–97.

Chapter 2

1. Jonathan Bloch is a Bighorn Basin researcher, and curator of vertebrate paleontology at the Florida Museum of Natural History.
2. Based on information in Winguth, A., et al., 2010, "Climate Response at the Paleocene-Eocene Thermal Maximum to Greenhouse Gas Forcing: A Model Study with CCSM3," *Journal of Climate*, 23, 2562–84.
3. Ansgar Walk, n.d., "Metasequoia Occidentalis Mummified Forest 1997-08-03.jpg," *Wikimedia Commons*.
4. The Gulf Stream water that reaches the far northern Atlantic is the densest water anywhere in the oceans because it is both salty and cold. It is denser than the water underneath, and so it sinks to become deep-ocean water.
5. Thompson, S., and Barron, E., 1981, "Comparison of Cretaceous and Present Earth Albedos: Implications for the Causes of Paleoclimates," *Journal of Geology*, 89(2), 143–67; Barron, E., Sloan, J., and Harrison, C., 1980, "Potential Significance of Land-sea Distribution and Surface Albedo Variations as a Climatic Forcing Factor: 180 m.y. to the Present," *Paleogeography, Paleoclimatology, Paleoecology*, 30, 17–40.
6. The latitude of approximately 65° north (or south) is ideal for glaciers to start forming and to grow because it has the appropriate balance of snowy winters and cool summers. Higher and colder latitudes (north or south) tend to have less snow, and lower latitudes have more summertime melting. In the context of the current distribution of continents, 65° north is good for glacial growth because there a lot of land at that latitude (65° north passes through Alaska,

Arctic Canada, Greenland, Iceland, Scandinavia, and Siberia), while 65° south is not because there is virtually no land at that latitude.

7. If you want to know more about how and why the Earth's climate has changed, you might like to read: Earle, S., 2021, *A Brief History of Earth's Climate*, New Society Publishers.

8. Smith, T., Rose, K., and Gingerich, P., 2006, "Rapid Asia-Europe-North America Geographical Dispersal of Earliest Eocene Primate *Teilhardina* During the Paleocene-Eocene Thermal Maximum," *Proceedings National Academy of Sciences*, 103, 11223–27.

9. Morse, P., et al., 2019, "New Fossils, Systematics, and Biogeography of the Oldest Known Crown Primate *Teilhardina* from the Earliest Eocene of Asia, Europe, and North America," *Journal of Human Evolution*, 128, 103–31.

10. During the Paleocene, there was significant volcanism in the northern Atlantic, partly because of the divergent plate boundary that created the Atlantic in the first place, and also because of the Iceland mantle plume (a plume of hot rock rising through the mantle, which still exists today, and is responsible for the existence of Iceland). The heat associated with these volcanic sources promoted expansion and therefore buoyancy of the upper mantle and crust, and that may have produced enough land to allow for a land route from Europe to North America.

Chapter 3

1. Robert Ballard is a retired Navy officer and a professor of oceanography at the University of Rhode Island.

2. Protists are microscopic single-celled organisms that are more complex than bacteria but cannot be classified as plants, fungi, or animals.

3. *Wikimedia*, "Emiliania huxleyi."

4. *Wikipedia*, by Hannes Grobe, 2011, "Foram globigerina."

5. The formula for calcium carbonate dissolution is as follows: $CaCO_3(s) + 2H^+ \longrightarrow Ca^{2+} + CO_2 + H_2O$. Increasing acidity adds H+ ions, and that forces the reaction to the right, leading to more dissolution.

6. Heezen, Bruce C., Tharp, Marie, and Ewing, Maurice, 1959, *The Floors of the Oceans: I. The North Atlantic,* Geological Society of America Special Papers, Vol. 65.

7. Hannes Grobe, 2009, "Joides-resolution ice odp.jpg," *Wikimedia Commons.*

8. An isotope is a version of an element that has a particular atomic weight because it has a different number of neutrons than other versions of the same element. The number of protons does not vary from one isotope of an element to another because it is the number of protons that determines what the element is. An oxygen atom cannot have 10 protons (it has to have 8), but it can have 10 neutrons.

9. For background on this relationship, see, for example, Barras, C., et al., 2010, "Calibration of $\delta^{18}O$ of Cultured Benthic Foraminiferal Calcite as a Function of Temperature," *Biogeosciences, European Geosciences Union,* 7(4), 1349–56.

10. Zachos, J., et al., 2001, "Trends, Rhythms, and Aberrations in Global Climate 65 Ma to Present," *Science,* 292, 686–93.

11. Westerhold, T., et al., 2020, "An Astronomically Dated Record of Earth's Climate and Its Predictability Over the Last 66 Million Years," *Science,* 369, 1383–87.

12. Kennett, J., and Stott, L., 1991, "Abrupt Deep-sea Warming, Palaeoceanographic Changes and Benthic Extinctions at the End of the Palaeocene," *Nature,* 225–29.

13. Based on information in Zachos, J., et al., 2006, "Extreme Warming of Mid-latitude Coastal Ocean During the Paleocene-Eocene Thermal Maximum: Inferences from TEX86 and Isotope Data," *Geology,* 34, 737–40. Figure 3.9 is also based on information in this source.

14. Trumbore, S., and Druffel, E., 1995, "Carbon Isotopes for Characterizing Sources and Turnover of Nonliving Organic Matter." *In Role of Nonliving Organic Matter in the Earth's Carbon Cycle,* John Wiley & Sons, 1995.

15. This is happening right now with anthropogenic climate change, and the changes are starting to affect the ability of some marine organisms to make shells.

16. Based on information in Shipboard Scientific Party, 2003, Leg 208 Preliminary Report. *ODP Prelim. Rpt.*, 108 [Online], www-odp. tamu.edu/publications/prelim/208_prel/208PREL.PDF

17. From FAQ 3.3 in: IPCC, 2021, *Climate Change 2021: The Physical Science Basis*, Contribution of Working Group I to the Sixth Assessment Report of the Intergovernmental Panel on Climate Change, Masson-Delmotte, V., et al., eds., Cambridge University Press.

18. IPCC, 2007, *Climate Change 2007: The Physical Science Basis*, Contribution of Working Group I to the Fourth Assessment Report of the Intergovernmental Panel on Climate Change, Solomon, S., et al., eds., Cambridge University Press, Cambridge, UK, and New York. Used with implicit permission from IPCC.

Chapter 4

1. Renowned marine biologist and oceanographer Sylvia Earle (no relation to the author), at a TED Talk in 2009.

2. Kennett, J., and Stott, L., 1991, "Abrupt Deep-sea Warming, Palaeoceanographic Changes and Benthic Extinctions at the End of the Palaeocene," *Nature*, 353, 225–29.

3. Based on Dunkley-Jones, T., et al., 2013, "Climate Model and Proxy Data Constraints on Ocean Warming Across the Paleocene–Eocene Thermal Maximum," *Earth-Science Reviews*, 125, 123–45.

4. Based on Nunes, F., and Norris, D., 2006, "Abrupt Reversal in Overturning During the Paleocene/Eocene Warm Period," *Nature*, 439, 60–63; Penman, D., and Turner, S., 2019, "The Influence of Circulation Change on Sedimentary Records of the Paleocene Eocene Thermal Maximum," *Pages Magazine*, 27(2).

5. Ilyina, T., and Heinze, M., 2019, "Carbonate Dissolution Enhanced by Ocean Stagnation and Respiration at the Onset of the Paleocene-Eocene Thermal Maximum," *Geophysical Research Letters*, 46, 842–52.

6. A rise of about 30 meters is suggested by Skye, Y., et al., 2021, "Shallow Marine Ecosystem Collapse and Recovery During the Paleocene-Eocene Thermal Maximum," *Global and Planetary*

Change, 207, 103649. A rise in the order of 20 meters is suggested by Speijer, R.P., and Morsi, A.M., 2002, "Ostracode Turnover and Sea-level Changes Associated with the Paleocene-Eocene Thermal Maximum," *Geology,* 2002, 30(1), 23–26.

7. Based on Sluijs, A., et al., 2014, "Warming, Euxinia and Sea Level Rise During the Paleocene–Eocene Thermal Maximum on the Gulf Coastal Plain: Implications for Ocean Oxygenation and Nutrient Cycling," *Climates Past,* 10, 1421–39. Figures 4.3 and 4.4 are based on this source.

8. Ilyina, T., and Heinze, M., 2019, *Geophysical Research Letters,* 46, 842–52.

9. As described in Chapter 3, the carbonate compensation depth is the water depth below which calcite becomes insoluble. Sediments that accumulate below that depth tend to have very little or no calcite; most are dominated by clay.

10. Based on data in Zachos, J., et al., 2005, "Rapid Acidification of the Ocean During the PETM," *Science,* 308, 1611–15.

11. Penman, D., et al., 2014, "Rapid and Sustained Surface Ocean Acidification During the PETM," *Paleoceanography,* 29; Babila, T., et al., 2022, "Surface Ocean Warming and Acidification Driven by Rapid Carbon Release Precedes the PETM," *Science Advances,* 8.

12. Ilyina and Heinze, 2019.

13. Kennett and Stott, 1991; Zachos, J., Dickens, Gerald R., and Zeebe, Richard E., 2008, "An Early Cenozoic Perspective on Greenhouse Warming and Carbon-Cycle Dynamics," *Nature,* 451, 279–283; Nunes and Norris, 2006.

14. Wright, J., and Schaller, M.F., 2013, "Evidence for a Rapid Release of Carbon at the PETM," *Proceedings National Academy of Sciences,* 110, 15908–13.

15. Alegret, L., Ortiz, S., and Molina, E., 2008, "Extinction and Recovery of Benthic Foraminifera Across the PETM at Alamedilla Section (Southern Spain)," *Paleogeography, Paleoclimatology Paleoecology,* 279(3–4), 186–200; Sluijs, A., et al., 2007, "The PETM Super Greenhouse: Biotic and Geochemical Signatures, Age Models and Mechanisms of Global Change." In Williams, M., et al., eds.,

Deep Time Perspectives on Climate Change: Marrying the Signal from Computer Models and Biological Proxies, Geological Society, London, 323–49; Speijer, R., et al., 2012, "Response of Marine Ecosystems to Deep-Time Global Warming: A Synthesis of Biotic Patterns Across the PETM," *Austrian Journal Earth Science,* 105, 6–16; Scheibner, C. and Speijer, R., 2007, "Decline of Coral Reefs During Late Paleocene to Early Eocene Global Warming, *eEarth Discussions,* 2, 133–50.

16. Aguirre, J., et al., 2022, "Coralline Algae at the Paleocene/Eocene Thermal Maximum in the Southern Pyrenees (N. Spain)," *Frontiers in Marine Science,* 9; Sheibner, C., and Speijer, R., 2007, Late Paleocene – Early Eocene Tethyan Carbonate Platform Evolution: A Response to Long- and Short-term Paleoclimactic Change," *Earth Science Reviews,* 90, 71–102.

17. Extirpation refers to the loss of a specific organism from a region, while the same organism continues to live in other regions.

18. See, for example, Prica, I., 2001, "Coralgal Facies of the Upper Eocene – Lower Oligocene Limestones in Letca-Rastci Area," *Studia Universitatis Babes-Bolyai, Geologica,* V, XLVI, 53–61.

19. Based on an illustration in Sheibner and Speijer, 2007.

20. Speijer et al., 2012.

Chapter 5

1. Gingerich, P., 2006, "Environment and Evolution Through the Paleocene-Eocene Thermal Maximum," *Trends in Ecology and Evolution,* 21, 246–53.

2. Based on information in Korasidis, V., et al., 2022, "Global Changes in Terrestrial Vegetation and Continental Climate During the Paleocene-Eocene Thermal Maximum," *Paleoceanography and Paleoclimatology,* 37; McInerney, F., and Wing, S., 2011, "The Paleocene-Eocene Thermal Maximum: A Perturbation of Carbon Cycle, Climate, and Biosphere with Implications for the Future," *Annual Reviews Earth Planetary Science,* 39, 489–516; and various other sources.

3. Based on information in Tierny, J., et al., 2022, "Spatial Patterns of Climate Change Across the PETM," *Proceedings of the National Academy of Sciences,* 119(42).

4. The Köppen Climate Classification is explained in some detail in Wikipedia.

5. Korasidis, V., et al., 2022. Figures 5.3 and 5.4 are based on maps from this source.

6. Schmitz, B., and Pujalte, V., 2007, "Abrupt Change in Seasonal Extreme Precipitation at the Paleocene-Eocene Boundary," *Geology*, 35, 215–18; Pujalte, V., Schmitz, B., and Payros, A., 2022, "A Rapid Sedimentary Response to the PETM Hydrological Change: New Data from Alluvial Units of the Tremp-Graus Basins (Spanish Pyrenees)," *Palaeogeography, Palaeoclimatology, Palaeoecology*, 589.

7. See, for example, Vimpere, L., et al., 2023, "Carbon Isotope and Biostratigraphic Evidence for an Expanded PETM Sedimentary Record in the Deep Gulf of Mexico," *Geology*, 51(4), 334–39.

8. Emanuel, K., 2005, "Increasing Destructiveness of Tropical Cyclones Over the Past 30 Years," *Nature*, 436, 686–88; Webster, P., et al., 2005, "Changes in Tropical Cyclone Number, Duration, and Intensity in a Warming Environment," *Science*, 309, 1844–46.

9. Kraus, M., et al., 2013, "Paleohydrologic Response To Continental Warming During the PETM," Bighorn Basin, Wyoming," *Palaeogeography, Palaeoclimatology, Palaeoecology*, 370; Wing, S., et al., 2009, "Coordinated Sedimentary and Biotic Change During the Paleocene-Eocene Thermal Maximum in the Bighorn Basin, Wyoming, USA." In *Climatic and Biotic Events of the Paleogene (CBEP 2009)*, Crouch, E., et al., eds., GNS Science Miscellaneous Series (no. 18), Institute of Geological and Nuclear Sciences, Lower Hutt, New Zealand.

10. Clyde, W.C., et al., 2013, "Bighorn Basin Coring Project (BBCP): A Continental Perspective on Early Paleogene Hyperthermals," *Scientific Drilling*, 16, 21–31.

11. Based on data in Bowen, G., et al., 2015, "Two Massive, Rapid Releases of Carbon During the Onset of the Palaeocene–Eocene Thermal Maximum," *Nature Geoscience*, 8, 44–47.

12. Ibid.

13. Bains, S., et al., 2003, "Marine-terrestrial Linkages at the Paleocene-Eocene Boundary." In: Wing, S., et al., eds., *Causes and Consequences of Globally Warm Climates in the Early Paleogene*, Geological Society

of American, Special Paper, 369, 1–9; Zhang, Q., et al., 2017, "Structure and Magnitude of the Carbon Isotope Excursion During the PETM," *Gondwana Research*, 46, 114–23.

14. Zachos, J., et al., 2006, "Extreme Warming of Mid-latitude Coastal Ocean During the PETM: Inferences from TEX86 and Isotope Data," *Geology*, 34, 737–40; Zachos, J., et al., 2005, "Rapid Acidification of the Ocean During the PETM, *Science*, 308, 1611–15.

15. Kennett, J., and Stott, L., 1991, "Abrupt Deep-sea Warming, Palaeoceanographic Changes and Benthic Extinctions at the End of the Palaeocene," *Nature*, 353, 225–29.

16. See, for example, Gingerich, P., 1975, "New North American Plesiadapidae (Mammalia, Primates) and a Biostratigraphic Zonation of the Middle and Upper Paleocene," *Contributions from the Museum of Paleontology*, University of Michigan, 24, 135–148, although discoveries of significant mammal fossils in the basin go back at least as far as Matthew, W., 1915, "A Revision of the Lower Eocene Wasatch and Wind River Faunas, Part IV, Entelonychia, Primates, Insectivores," *Bulletin of the American Museum of Natural History*, 34, 429–83.

17. Wing, S., and Currano, E., 2013, "Plant Response to a Global Greenhouse Event 56 Million Years Ago," *American Journal of Botany*, 100, 1234–54.

18. Wing, S., and Hickey, L., 1984, "The Platycarya Perplex and the Evolution of the Juglandaceae," *American Journal of Botany*, 71, 388–411.

19. Handley, L., Crouch, E.M., and Pancost, R.D., 2011, "A New Zealand Record of Sea Level Rise and Environmental Change During the PETM," *Palaeogeography, Palaeoclimatology, Palaeoecology*, 305, 185–200.

20. Jaramillo, C., et al., 2010, "Effects of Rapid Global Warming at the Paleocene-Eocene Boundary on Neotropical Vegetation, *Science*, 330, 957–61.

21. Schmitz, B., Pujalte, V., and Núñez-Betelu, K., 2001, "Climate and Sea-level Perturbations During the Initial Eocene Thermal

Maximum: Evidence from Siliciclastic Units in Basque Basin (Ermua, Zumaia and Trabuka Pass), Northern Spain," *Palaeogeography, Palaeoclimatology, Palaeoecology,* 165, 299–320.

22. Schmitz, B., and Pujalte, V., 2007; Pujalte, V., et al., 2022.
23. Kender, S., et al., 2012, "Marine and Terrestrial Environmental Changes in NW Europe Preceding Carbon Release at the Paleocene-Eocene Transition," *Earth and Planetary Science Letters,* 353, 108–20.
24. Sluijs, A., et al., 2008, "Arctic Late Paleocene-Early Eocene Paleo-environments with Special Emphasis on the PETM (Lomonosov Ridge, Integrated Ocean Drilling Program Expedition 302)," *Paleoceanography,* 23(1).
25. Gingerich, P., 2006.
26. Based on data in Fraser, D., and Lyons, S., 2020, "Mammal Community Structure Through the PETM," *American Naturalist,* 196.
27. Gingerich, P., 2006.
28. Bown, T., and Gingerich, P., 1972, "Dentition of the Early Eocene Primates *Niptomomys* and *Absarokius,*" *Postilla,* 158, 1–10; Ramos, E., et al., 2022, "Swift Weathering Response on Floodplains During the PETM," *Geophysical Research Letters,* 49, 1–10.
29. Rose, K., 1981, "The Clarkforkian Land-mammal Age and Mammalian Faunal Composition Across the Paleocene-Eocene Boundary," University of Michigan, *Papers on Paleontology,* 26.
30. Gingerich, P., 2003, *Geological Society of America Special Paper* 369, 463–78.
31. Based on data in Vitek, N., et. al, 2021, "Evaluating the Responses of Three Closely Related Small Mammal Lineages to Climate Change Across the PETM," *Paleobiology,* 47(3), 464–86.
32. Smith, J., et al., 2009, "Transient Dwarfism of Soil Fauna During the PETM," *Proceedings National Academy of Sciences,* 106(42), 17655–60.
33. Bergmann, C., 1847, "Über die Verhältnisse der Wärmeökonomie der Thiere zu ihrer Grösse," *Göttinger Studien,* 1, 595–708.
34. Gingerich, P., 2003.

Chapter 6

1. Gavin Schmidt is a climate modeler working for NASA and at Columbia University. The statement was made in 2009, and although we now know a great deal more about the PETM, there is still uncertainty about its specific cause.
2. Penman, D., and Zachos, J., 2018, "New Constraints on Massive Carbon Release and Recovery Processes During the PETM," *Environmental Research Letters*, 13(10).
3. Ibid.; Ridgwell, A., 2007, "Interpreting Transient Carbonate Compensation Depth Changes by Marine Sediment Core Modeling," *Paleoceanography and Paleoclimatology*, 22(4).
4. DeConto, R., et al., 2010, "Hyperthermals and Orbitally Paced Permafrost Soil Organic Carbon Dynamics." Presented at American Geophysical Union Fall Meeting, San Francisco. Most permafrost on the Earth now exists in areas that were glaciated until as recently as a few thousand to 15,000 years ago. Permafrost does not typically form at the base of icesheets because the temperature is close to or above freezing, and so there has only been several thousand years for permafrost to accumulate in deglaciated regions. Even if permafrost could form at the base of a glacier, there isn't enough biological activity in that setting to allow methane to be produced and accumulate.
5. DeConto, R., et al., 2012, "Past Extreme Warming Events Linked to Massive Carbon Release from Thawing Permafrost," *Nature*, 484, 87–91.
6. Rampino, M., 2013, "Peraluminous Igneous Rocks as an Indicator of Thermogenic Methane Release from the North Atlantic Volcanic Province at the Time of the PETM," *Bulletin of Volcanology*, 75, 1–5.
7. Saunders, A., et al., 1997, "The North Atlantic Igneous Province." In Mahoney, J.J., and Coffin, M.F., eds., *Large Igneous Provinces: Continental, Oceanic and Planetary Flood Volcanism*, Washington DC, American Geophysical Union, Monograph 100, 45–93.
8. Frieling, J., et al., 2016, "Thermogenic Methane Release as a Cause for the Long Duration of the PETM," *Proceedings National Academy of Science*, 113, 12059–64; Stokke, E., et al., 2020, "Rapid

and Sustained Environmental Responses to Global Warming: The PETM in the Eastern North Sea," *Climate of the Past,* 7(5),1989–2013; Schmitz, B., et al., 2004, "Basaltic Explosive Volcanism, but No Comet Impact, at the Paleocene–Eocene Boundary: High-resolution Chemical and Isotopic Records from Egypt, Spain and Denmark," *Earth and Planetary Science Letters,* 225, 1–17.

9. Berndt, C., et al., 2023, "Shallow-water Hydrothermal Venting Linked to the PETM," *Nature Geoscience,* 16, 803–809.

10. Based on Stokke, E., et al., 2020.

11. This calculation is based on the assumption of a 180,000-year-long PETM and that the immediate pre-PETM rate of sedimentation was similar to that during the PETM. The latter assumption is questionable, and the time estimate could easily be wrong by a factor of two. At other marine locations, the rate of sedimentation dropped during the PETM, so it is more likely that the interval is less than 1500 y, rather than more than.

12. Based on Zeebe, R., and Lourens, L., 2019, "Solar System Chaos and the Paleocene-Eocene Boundary Age Constrained by Geology and Astronomy," *Science,* 365, 926–29.

13. Cramer, B., and Kent, D., 2005, "Bolide Summer: The PETM as a Response to an Extraterrestrial Trigger," *Palaeogeography, Palaeoclimatology, Palaeoecology,* 224, 144–66; Kent, D., et al., 2003, "A Case for a Comet Impact Trigger for the PETM and Carbon Isotope Excursion," *Earth and Planetary Science Letters,* 211, 13–26.

14. The element iridium is significantly enriched in comets and meteoroids compared with most rocks of the Earth's crust. There is a famous example of iridium enrichment associated with the end-Cretaceous impact that killed the dinosaurs. Iridium is also weakly enriched in volcanic rocks and ash relative to other crustal rocks.

15. Schmitz, B., et al., 2004.

16. Based on data in Zachos, J., et al., 2005, "Rapid Acidification of the Ocean During the PETM," *Science,* 308, 1611–15; Bowen, G., et al., 2015, "Two Massive Rapid Releases of Carbon During the Onset of the Palaeocene–Eocene Thermal Maximum, *Nature Geoscience,* 8, 44–47.

17. It is likely that the methane that was released to produce the CIE shown in figure 6.5 had a ^{13}C level of -3% or lower. The recorded excursion is much less than that (-0.6% ^{13}C) because the methane released was mixed into an atmosphere made up of carbon with a ^{13}C level close to 0.

18. Zhu, M., et al., 2010, "High-resolution Carbon Isotope Record for the PETM from the Nanyang Basin, Central China," *Chinese Science Bulletin*, 55, 3606–11.

19. Kirtland-Turner, S., 2018, "Constraints on the Onset Duration of the PETM," *Philosophical Transactions of the Royal Society*, A, 376.

20. Bains, S., et al., 2000, "Termination of Global Warmth at the Palaeocene/Eocene Boundary Through Productivity Feedback," *Nature*, 407, 171–74.

21. Torfstein, A., et al., 2010, "Productivity Feedback Did Not Terminate the PETM," *Climates Past*, 6, 265–72.

Chapter 7

1. Peter Wadhams is emeritus professor of Ocean Physics and head of the Polar Ocean Physics Group in the Department of Applied Mathematics and Theoretical Physics at University of Cambridge. Speaking in 2012, he was referring to the consequences of the loss of Arctic sea ice.

2. Pistone, K., Eisenman, I., and Ramanathan, V., 2014, "Observational Determination of Albedo Decrease Caused by Vanishing Arctic Sea Ice," *Proceedings of the National Academy of Science*, 111(9), 3322–26.

3. Schweiger, A., et al., 2011, "Uncertainty in Modeled Arctic Sea Ice Volume," *Journal of Geophysical Research*, 116(C8).

4. Sea ice starts melting in May of each year, and it continues to melt through the summer, with maximum ice loss reached in September. New ice starts to grow in October, and it reaches its maximum area in April.

5. Riihelä, A., Bright, R.M., and Anttila, K., 2021, "Recent Strengthening of Snow and Ice Albedo Feedback Driven by Antarctic Sea-Ice Loss," *Nature Geoscience*, 14, 832–36.

6. Australian Antarctic Program Partnership, "Deep Freeze Falters: Antarctic Sea Ice Drops to New Low," https://aappartnership.org.au, February 27, 2023.
7. National Snow and Ice Data Center, "Antarctic Sets a Record Low Maximum by Wide Margin," http://nsidc.org, September 2023.
8. IPCC, 2021, "Summary for Policymakers." In *Climate Change 2021: The Physical Science Basis.* Contribution of Working Group I to the Sixth Assessment Report of the Intergovernmental Panel on Climate Change, Masson-Delmotte, V., et al., eds., Cambridge University Press.
9. Turetsky, M., et al., 2019, "Permafrost Collapse Is Accelerating Carbon Release," *Nature*, 569(7754), 32–34.
10 Based on Rounce, D., et al., 2023, "Global Glacier Change in the 21st Century: Every Increase in Temperature Matters," *Science,* 379(6627), 78–83.
11. Biskaborn, B., et al., 2019, "Permafrost Is Warming at a Global Scale," *Nature Communications,* 10, Article 264.
12. Li, Q., et al., 2023, "Abyssal Ocean Overturning Slowdown and Warming Driven by Antarctic Meltwater," *Nature*, 615, 841–47.
13. Readfearn, G., "Melting Antarctic Ice Predicted to Cause Rapid Slowdown of Deep-Ocean Current by 2050," *The Guardian*, March 30, 2023.
14. Hood, E., et al., 2015, "Storage and Release of Organic Carbon from Glaciers and Ice Sheets," *Nature Geoscience*, 8, 91–96.
15. Based on data in Friedlingstein, P., et al., 2022, "Global Carbon Budget 2022," *Earth System Sciences Data*, 14, 4811–900. The dashed lines extending from 2021 to 2120 are possible scenarios for future emissions, and are based on a 2022 diagram by H. Ritchie and co-authors at https://ourworldindata.org/co2-and-other-greenhouse-gas-emissions
16. Based on data compiled by Ritchie, H., and Roser, M., 2021, "Forests and Deforestation," https://ourworldindata.org
17. Black, B., et al., 2018, "Systemic Swings in End-Permian Climate from Siberian Traps Carbon and Sulfur Outgassing," *Nature Geoscience*, 11, 949–54.

Chapter 8

1. Arias, P., et al., 2021, *Climate Change 2021: The Physical Science Basis. Contribution of Working Group I to the IPCC Sixth Assessment Report of the Intergovernmental Panel on Climate Change,* Masson-Delmotte, V., et al., eds., Cambridge University Press.

2. Knutson, T., et al., 2015, "Global Projections of Intense Tropical Cyclone Activity for the Late Twenty-First Century from Dynamical Downscaling of CMIP5/RCP4.5 Scenarios," *Journal of Climate,* 28, 7203–24.

3. The Reef-World Foundation, March 1, 2021, "What Would Happen If There Were No Coral Reefs," https://reef-world.org

4. Kieffer, M., n.d., "Staghorn coral," *Wikimedia Commons.*

5. Bryndum-Bucholz, A., et al., 2018, "21st Century Climate Change Impacts on Marine Animal Structure Across Ocean Biomass and Ecosystem Basins," *Global Change Biology,* 25(2), 459–72.

6. Mekkes, L., et al., 2021, "Effects of Ocean Acidification on the Sub-Antarctic Pteropod *Limacina Retroversa,*" *Frontiers in Marine Science,* 8, Article 581432.

7. Ross Hopcroft, 2005, "Limacina helicina," no.7, *Wikimedia Commons.*

8. Graham, A., et al., 2022, "Rapid Retreat of the Thwaites Glacier in the Pre-satellite Era," *Nature Geoscience,* 15, 706–13.

9. Boers, N., 2021, "Observation-based Early-warning Signals for a Collapse of the Atlantic Meridional Overturning Circulation," *Nature Climate Change,* 11, 680–88; Ditlevsen, P., and Ditlevsen, S., 2023, "Warning of a Forthcoming Collapse of the Atlantic Meridional Overturning Circulation," *Nature Communications,* 14 (4254).

10. Li, Q., et al., 2023, "Abyssal Ocean Overturning Slowdown and Warming Driven by Antarctic Meltwater," *Nature,* 615, 841–47.

11. Shivaram, D., "Heat Wave Killed an Estimated 1 Billion Sea Creatures, and Scientists Fear Even Worse," National Public Radio, www.npr.org, July 9, 2021.

12. By bisonlux, www.flickr.com/photos/26912057@N02/2525803211.

13. Vitousek, S., et al., 2023, "A Model Integrating Satellite-Derived Shoreline Observations for Predicting Fine-scale Shoreline Response

to Waves and Sea-level Rise Across Large Coastal Regions," ESS Open Archive, March 09, 2023.

Chapter 9

1. Arias, P., et al., 2021, *Climate Change 2021: The Physical Science Basis.* Contribution of Working Group I to the IPCC Sixth Assessment Report of the Intergovernmental Panel on Climate Change, Masson-Delmotte, V., et al., eds., Cambridge University Press.
2. NASA, https://data.giss.nasa.gov/gistemp/graphs_v4
3. Dai, A., 2012, "Increasing Drought Under Global Warming in Observations and Models," *Nature Climate Change*, 3, 52–58.
4. Based on Hausfather, Z., 2018, "Explainer: What Climate Models Tell Us About Future Rainfall," www.carbonbrief.org
5. Tyukavina, A., et al., 2022, "Global Trends of Forest Loss Due to Fire from 2001 to 2019," *Frontiers in Remote Sensing*, 3; McCarthy, J., et al., 2022, "New Data Confirms: Forest Fires Are Getting Worse," *World Resources Institute*, updated August 2023, www.wri.org
6. Whitman, E., et al., 2019, "Short-interval Wildfire and Drought Overwhelm Boreal Forest Resilience," *Nature Scientific Reports*, 9, Article 18796.
7. Fischer, E., et al., 2014, "Models Agree on Forced Response Pattern of Precipitation and Temperature Extremes," *Geophysical Research Letters*, 41(23), 8554–62.
8. Pulliainen, J., et al., 2020, "Patterns and Trends of Northern Hemisphere Snow Mass from 1980 to 2018," *Nature*, 581, 294–301.
9. Parmesan, C., 2006, "Ecological and Evolutionary Responses to Recent Climate Change," *Annual Review Ecology Evolution and Systematics*, 37, 637–69.
10. Arias, P., et al., 2021.
11. Lenton, T., et al., May, 22, 2023, "Quantifying the Human Cost of Global Warming," *Nature Sustainability*, www.nature.com/articles
12. Vecellio, D., et al., 2022, "Evaluating the 35°C Wet-bulb Temperature Adaptability Threshold for Young, Health Subjects (PSU HEAT Project)," *Journal of Applied Physiology*, 132, 340–45.

13. *Extreme Heat and Human Mortality: A Review of Heat-Related Deaths in B.C. in Summer 2021*, Report to the Chief Coroner of British Columbia, June 7, 2022, www2.gov.bc.ca

14. Robinson, K, 2020, *The Ministry for the Future*, Hachette, New York.

15. Lenton, T., et al., 2023.

16. Figure 9.5 is based on diagram in Xu, C., et al., 2019, "Future of the Human Climate Niche," *Proceedings of the National Academy of Science*, 117, 11350–55.

17. Mankin, J., et al., 2015, "The Potential for Snow to Supply Human Water Demand in the Present and Future," *Environmental Research Letters*, 10, 114016.

18. World Bank, https://datacatalog.worldbank.org/search/dataset/00 41449

19. Kulp, S., and Srauss, B., 2019, "New Elevation Data Triple Estimates of Global Vulnerability to Sea-level Rise and Coastal Flooding," *Nature Communications*, 10, Article number 4844.

20. Dayan, H., et al., 2021, "High-end Scenarios of Sea-level Rise for Coastal Risk-averse Stakeholders," *Frontiers in Marine Science*, 8.

21. There are currently 113 million people living in the just the 20 most populus cities at risk by 2070. Of those, 3 are in North America, 3 are in Africa, and 14 are in Asia. Nichols, R., et al., 2007, "Ranking of the World's Cities Most Exposed to Coastal Flooding Today and in the Future," OECD Environment Working Paper No. 1, www.oecd.org

22. Based on Poore, R., Williams, R.S. Jr., and Tracey, C., 2000, "Sea Level and Climate," US Geological Survey Fact Sheet 002–00, http://pubs.usgs.gov

23. Mbow, C., et al., 2019, "Food Security." *In Climate Change and Land*, Shukla, P.R., et al., eds., an IPCC Special Report on climate change, desertification, land degradation, sustainable land management, food security, and greenhouse gas fluxes in terrestrial ecosystems. Available at www.ipcc.ch/site/assets/uploads/sites/4/2022/11/SRC CL_Chapter_5.pdf

24. Ebi, K., and Loladze, I., 2019, "Elevated Atmospheric CO_2 Concentrations and Climate Change Will Affect Our Food's Quality and Quantity," *Lancet Planetary Health*, 3, E283–84.

Chapter 10

1. Vince, G., 2022, *Nomad Century: How Climate Migration Will Reshape Our World*, Flatiron, New York.
2. Based on data in United Nations High Commissioner for Refugees, 2022, *Global Trends Report 2022*, www.unhcr.org
3. Based on data in International Displacement Monitoring Centre, 2023, *2023 Global Report on Internal Displacement*, IDMC, Geneva.
4. Based on data in 2023 *Global Report on Internal Displacement.*
5. Andy Hall, 2011, from "Oxfam East Africa: A Mass Grave for Children in Dadaab," *Wikimedia Commons.*
6. International Displacement Monitoring Centre, 2023.
7. *Wikipedia*, 2023, "2020–2023 Horn of Africa Drought."
8. Based on information in Kimutai, J., et al., 2023, *Human-induced Climate Change Increased Drought Severity in Horn of Africa*, Imperial College London, https://doi.org/10.25561/103482
9. Ibid.
10. Lenton, T., et al., 2023, "Quantifying the Human Cost of Global Warming," *Nature Sustainability*, www.nature.com/articles; Xu, C., et al., 2019, "Future of the Human Climate Niche," *Proceedings of the National Academy of Science*, 117, 11350–55.
11. Otte, J., "Florida Rocked by Home Insurance Crisis: 'I May Have to Sell Up and Leave,'" *The Guardian*, July 15, 2023.
12. United Nations Framework Convention on Climate Change, *Paris Agreement*, https://unfccc.int/sites/default/files/english_paris_agreement.pdf
13. Vince, G., 2022.

Chapter 11

1. James Lovelock (1919–2022) was an English scientist and environmentalist who is best known for proposing the Gaia hypothesis.
2. Based on figure TS20 in IPCC, 2021, *Climate Change 2021: The Physical Science Basis*. Contribution of Working Group I to the Sixth Assessment Report of the Intergovernmental Panel on Climate Change, Masson-Delmotte, V., et al., eds., Cambridge University Press.

3. Tollefson, J., 2021, "COVID Curbed 2020 Carbon Emissions—But Not by Much," *Nature*, 589, 343.

4. Venter, Z., et al., 2020, "COVID-19 Lockdowns Cause Global Air Pollution Declines," *Proceedings of the National Academy of Science*, 117, 18984–90.

5. Based on data in Chen, K., et al., 2020, "Air Pollution Reduction and Mortality Benefit During the COVID-19 Outbreak in China," *Lancet Planet Health*, 4, e210–12.

6. Ibid.

7. Milner, J., et al., 2023, "Impact on Mortality of Pathways to Net Zero GHG Emissions in England and Wales: A Multisectoral Modelling Study," *Lancet Planetary Health*, 7(2), e128–e136.

8. Warren, T., 2023, "Payback Period for Solar Panels," Energy Bot. According to this article, the average payback time for home solar installations in the US is 10 years, although it is much less in some states. Solar installations typically have warranties of at least 20 years, and they will normally last longer than that.

9. Richi, E., et al., 2015, "Health Risks Associated with Meat Consumption: A Review of Epidemiological Studies," *International Journal of Vitamin and Nutrition Research*, 85, 70–78.

10. Based on information in Poore, J., and Nemecek, T., 2018, "Reducing Food's Environmental Impacts Through Producers and Consumers," *Science*, 360, 987–92.

11. Ibid. Also based on information from H. Ritchie at https://ourworldindata.org/food-emissions-carbon-budget

12. Ben-Shahar, O., 2016, "The Non-voters Who Decided the Election: Trump Won Because of Lower Democratic Turnout," *Forbes*, November 17, 2016.

13. Based on data in International Energy Agency, n.d., "Energy Subsidies: Tracking the Impact of Fossil-fuel Subsidies," www.iea.org

14. Based on data in Net Zero Tracker, 2021, Energy and Climate Intelligence Unit, Data-Driven EnviroLab, NewClimate Institute, Oxford Net Zero.

15. Greenfield, P., 2023, "Revealed: More than 90% of Rainforest Carbon Offsets by Biggest Certifier Are Worthless, Analysis Shows," *The Guardian,* January 18, 2023.

16. Much of the following is based on data and commentary compiled and written by Ritchie, H., Rosado, P., and Roser, M., from https://ourworldindata.org/carbon-pricing. The data are largely from Dolphin, G., et al., 2020, "The Political Economy of Carbon Pricing: A Panel Analysis," *Oxford Economic Papers,* 72, 472–500; Dolphin, G., and Xiahou, Q., 2022, "World Carbon Pricing Database: Sources and Methods," *Nature Scientific Data,* 9(573).

17. Venzke, I., 2023, "We Are Fucked" vs. "It's Not Too Late," *European Review of Books,* No. 3, 32–45.

18. See, for example, Long, H., 2019, "This Is Not Controversial: Bipartisan Group of Economists Calls for Carbon Tax," *Washington Post,* January 16, 2019; IPCC, 2022, *Climate Change 2022: Mitigation of Climate Change.* Contribution of Working Group III to the Sixth Assessment Report of the Intergovernmental Panel on Climate Change, Shukla, P., et al., eds., Cambridge University Press.

19. Wood, N., Lawlor, R., and Freear, J., 2023, "Rationing and Climate Change Mitigation," *Ethics, Policy and Environment,* online February 2023.

20. Milan, R., et al., 2017, "Closing the Induced Vehicle Travel Gap Between Research and Practice," *Transportation Research Record,* 2653(1), 10–16.

21. CBC News, July 15, 2019, "'They Were Saying No One Would Ride It': 10 Years On, Burrard Bike Lane Is N. America's Busiest, Officials Say," www.cbc.ca/news.

22. Simon, D., 2013, *Meatonomics,* Conari Press, Berkeley; Sewell, C., 2020, "Removing the Meat Subsidy: Our Cognitive Dissonance Around Animal Agriculture," *Journal of International Affairs,* February, 11, 2020.

Appendix

1. The actual chemical reaction is as follows: $CaAl_2Si_2O_8 + H_2CO_3 + \frac{1}{2}O_2 \longrightarrow Al_2Si_2O_5(OH)_4 + Ca^{2+} + CO_3^{2-}$. The feldspar shown

here ($CaAl_2Si_2O_8$) is calcium plagioclase (anorthite). There are also sodium (albite) and potassium (orthoclase) versions of feldspar.

2. Based on Wing, S., and Currano, E., 2013, "Plant Response to a Global Greenhouse Event 56 Million Years Ago," *American Journal of Botany*, 100, 1234–54; Korasidis, V., et al., 2022, "Biostratigraphically Significant Palynofloras from the Paleocene–Eocene Boundary of the USA," *Palynology*, 47(1).

3. Based on information in Gingerich, P., 2003, "Mammalian Responses to Climate Change at the Paleocene-Eocene Boundary: Polecat Bench Record in the Northern Bighorn Basin," *Geological Society of America Special Paper* 369, 463–78.

4. Based on Figure TS20 in IPCC, 2021, *Climate Change 2021: The Physical Science Basis.* Contribution of Working Group I to the Sixth Assessment Report of the Intergovernmental Panel on Climate Change, Masson-Delmotte, V., et al., eds., Cambridge University Press.

5. Based in part on a diagram by R. Rohde, "Atmospheric Transmission, for the Global Warming Art Project," *Wikimedia Commons*.

6. Based on data in Poore, J., and Nemecek, T., 2018, "Reducing Food's Environmental Impacts Through Producers and Consumers," *Science*, 360, 987–92.

7. There are numerous online carbon emission calculators. The best known is www.carbonfootprint.com/calculator.aspx, but others will do the job. The estimate from the simple one here should be generally comparable to the results from carbonfootprint.com

Index

About the Author

Steven Earle, PhD, has developed and taught university earth science courses for over four decades. He is the author of *A Brief History of Climate Change* and the widely used textbook *Physical Geology*. A dedicated community activist, he champions climate change solutions in areas such as human-powered and low-carbon transportation, home energy, and land stewardship.

ABOUT NEW SOCIETY PUBLISHERS

New Society Publishers is an activist, solutions-oriented publisher focused on publishing books to build a more just and sustainable future. Our books offer tips, tools, and insights from leading experts in a wide range of areas.

We're proud to hold to the highest environmental and social standards of any publisher in North America. When you buy New Society books, you are part of the solution!

- This book is printed on **100% post-consumer recycled paper,** processed chlorine-free, with low-VOC vegetable-based inks (since 2002).
- Our corporate structure is an innovative employee shareholder agreement, so we're one-third employee-owned (since 2015)
- We've created a Statement of Ethics (2021). The intent of this Statement is to act as a framework to guide our actions and facilitate feedback for continuous improvement of our work
- We're carbon-neutral (since 2006)
- We're certified as a B Corporation (since 2016)
- We're Signatories to the UN's Sustainable Development Goals (SDG) Publishers Compact (2020–2030, the Decade of Action)

At New Society Publishers, we care deeply about *what* we publish—but also about *how* we do business.

To download our full catalog, sign up for our quarterly newsletter, and learn more about New Society Publishers, please visit newsociety.com

ENVIRONMENTAL BENEFITS STATEMENT

New Society Publishers saved the following resources by printing the pages of this book on chlorine free paper made with 100% post-consumer waste.

TREES	WATER	ENERGY	SOLID WASTE	GREENHOUSE GASES
19	1,500	8	64	8,090
FULLY GROWN	GALLONS	MILLION BTUs	POUNDS	POUNDS

 Environmental impact estimates were made using the Environmental Paper Network Paper Calculator 4.0. For more information visit www.papercalculator.org

Certified
B Corporation

new society
PUBLISHERS
www.newsociety.com

FSC
www.fsc.org
MIX
Paper | Supporting responsible forestry
FSC® C016245

SDG PUBLISHERS COMPACT